AF785558

Starch - Evolution and Recent Advances

Edited by Martins Ochubiojo Emeje

Published in London, United Kingdom

Starch - Evolution and Recent Advances
http://dx.doi.org/10.5772/intechopen.94824
Edited by Martins Ochubiojo Emeje

Contributors
Ashoka Gamage, Thiviya Punniamoorthy, Terrence Madhujith, Teodoro Suarez-Diéguez, Juan Antonio Nieto, Youkabed Zarrouk, Mouna Boulares, Dorra Sfayhi, Bechir Slimi, Saleem Siddiqui, Naseer Ahmed, Neeraj Phogat, Veda Krishnan, Archana Singh, Monika Awana, Debarati Mondal, Piyush Verma, Shelly Praveen, Doina Dimonie, Laura-Nicoleta Dragomir, Mircea Filipescu, Alina_Elena Mustetea, Mihail Dragne, Georgina Calderón-Domínguez, Verónica Cuellar-Sánchez, Ma.Paz Salgado-Cruz, Arturo García-Borquez, Eduardo Morales-Sánchez, Alitzel B. García-Hernández, Reynold R. Farrera-Rebollo, Marcela González-Vázquez, Fatima Sarahi Serrano-Villa, Judith Eloyi John, Obi Peter Adigwe, Revati Wanikar, Swati Kotwal, Henry Omoregie O. Egharevba, Martins Ochubiojo Emeje

© The Editor(s) and the Author(s) 2022
The rights of the editor(s) and the author(s) have been asserted in accordance with the Copyright, Designs and Patents Act 1988. All rights to the book as a whole are reserved by INTECHOPEN LIMITED. The book as a whole (compilation) cannot be reproduced, distributed or used for commercial or non-commercial purposes without INTECHOPEN LIMITED's written permission. Enquiries concerning the use of the book should be directed to INTECHOPEN LIMITED rights and permissions department (permissions@intechopen.com).
Violations are liable to prosecution under the governing Copyright Law.

Individual chapters of this publication are distributed under the terms of the Creative Commons Attribution 3.0 Unported License which permits commercial use, distribution and reproduction of the individual chapters, provided the original author(s) and source publication are appropriately acknowledged. If so indicated, certain images may not be included under the Creative Commons license. In such cases users will need to obtain permission from the license holder to reproduce the material. More details and guidelines concerning content reuse and adaptation can be found at http://www.intechopen.com/copyright-policy.html.

Notice
Statements and opinions expressed in the chapters are these of the individual contributors and not necessarily those of the editors or publisher. No responsibility is accepted for the accuracy of information contained in the published chapters. The publisher assumes no responsibility for any damage or injury to persons or property arising out of the use of any materials, instructions, methods or ideas contained in the book.

First published in London, United Kingdom, 2022 by IntechOpen
IntechOpen is the global imprint of INTECHOPEN LIMITED, registered in England and Wales, registration number: 11086078, 5 Princes Gate Court, London, SW7 2QJ, United Kingdom
Printed in Croatia

British Library Cataloguing-in-Publication Data
A catalogue record for this book is available from the British Library

Additional hard and PDF copies can be obtained from orders@intechopen.com

Starch - Evolution and Recent Advances
Edited by Martins Ochubiojo Emeje
p. cm.

This title is part of the Biochemistry Book Series, Volume 33
Topic: Chemical Biology
Series Editor: Miroslav Blumenberg
Topic Editors: Şükrü Beydemir and Deniz Ekinci

Print ISBN 978-1-83969-890-3
Online ISBN 978-1-83969-891-0
eBook (PDF) ISBN 978-1-83969-892-7
ISSN 2632-0983

We are IntechOpen, the world's leading publisher of Open Access books

Built by scientists, for scientists

5,800+
Open access books available

144,000+
International authors and editors

180M+
Downloads

156
Countries delivered to

Our authors are among the
Top 1%
most cited scientists

12.2%
Contributors from top 500 universities

CLARIVATE ANALYTICS
BOOK CITATION INDEX
INDEXED

WEB OF SCIENCE™

Selection of our books indexed in the Book Citation Index
in Web of Science™ Core Collection (BKCI)

Interested in publishing with us?
Contact book.department@intechopen.com

Numbers displayed above are based on latest data collected.
For more information visit www.intechopen.com

For EU product safety concerns:
IN TECH d.o.o., Prolaz Marije Krucifikse Kozulić 3, 51000 Rijeka, Croatia,
info@intechopen.com or visit our website at intechopen.com.

IntechOpen Book Series
Biochemistry
Volume 33

Aims and Scope of the Series

Biochemistry, the study of chemical transformations occurring within living organisms, impacts all of the life sciences, from molecular crystallography and genetics, to ecology, medicine and population biology. Biochemistry studies macromolecules - proteins, nucleic acids, carbohydrates and lipids –their building blocks, structures, functions and interactions. Much of biochemistry is devoted to enzymes, proteins that catalyze chemical reactions, enzyme structures, mechanisms of action and their roles within cells. Biochemistry also studies small signaling molecules, coenzymes, inhibitors, vitamins and hormones, which play roles in the life process. Biochemical experimentation, besides coopting the methods of classical chemistry, e.g., chromatography, adopted new techniques, e.g., X-ray diffraction, electron microscopy, NMR, radioisotopes, and developed sophisticated microbial genetic tools, e.g., auxotroph mutants and their revertants, fermentation, etc. More recently, biochemistry embraced the 'big data' omics systems. Initial biochemical studies have been exclusively analytic: dissecting, purifying and examining individual components of a biological system; in exemplary words of Efraim Racker, (1913 –1991) "Don't waste clean thinking on dirty enzymes." Today, however, biochemistry is becoming more agglomerative and comprehensive, setting out to integrate and describe fully a particular biological system. The 'big data' metabolomics can define the complement of small molecules, e.g., in a soil or biofilm sample; proteomics can distinguish all the proteins comprising e.g., serum; metagenomics can identify all the genes in a complex environment e.g., the bovine rumen.
This Biochemistry Series will address both the current research on biomolecules, and the emerging trends with great promise.

Meet the Series Editor

Miroslav Blumenberg, Ph.D., was born in Subotica and received his BSc in Belgrade, Yugoslavia. He completed his Ph.D. at MIT in Organic Chemistry; he followed up his Ph.D. with two postdoctoral study periods at Stanford University. Since 1983, he has been a faculty member of the RO Perelman Department of Dermatology, NYU School of Medicine, where he is codirector of a training grant in cutaneous biology. Dr. Blumenberg's research is focused on the epidermis, expression of keratin genes, transcription profiling, keratinocyte differentiation, inflammatory diseases and cancers, and most recently the effects of the microbiome on the skin. He has published more than 100 peer-reviewed research articles and graduated numerous Ph.D. and postdoctoral students.

Meet the Volume Editor

Martins Emeje obtained a BPharm with distinction from Ahmadu Bello University, Nigeria, and an MPharm and Ph.D. from the University of Nigeria (UNN), where he received the best Ph.D. award and was enlisted as UNN's "Face of Research." He established the first nanomedicine center in Nigeria and was the pioneer head of the intellectual property and technology transfer as well as the technology innovation and support center. Prof. Emeje's several international fellowships include the prestigious Raman fellowship. He has published more than 150 articles and patents. He is also the head of R&D at NIPRD and holds a visiting professor position at Nnamdi Azikiwe University, Nigeria. He has a postgraduate certificate in Project Management from Walden University, Minnesota, as well as a professional teaching certificate and a World Bank certification in Public Procurement. Prof. Emeje was a national chairman of academic pharmacists in Nigeria and the 2021 winner of the May & Baker Nigeria Plc–sponsored prize for professional service in research and innovation.

Contents

Preface

That starch is the second most abundant biomass in nature, as well as a cheap, renewable, and biodegradable polymer is indisputable. Historically, starch finds wide use in the food, textiles, cosmetics, plastics, adhesives, paper, and pharmaceutical industries. Today, starch is either used in its native or modified form. It is estimated that the starch output value is about €15 billion per year. The leading users of starch are believed to be the brewing, baking powder, and confectionery industries. The need for starch continues to increase and advances in its applications continue to emerge. Starch and starch derivatives play very important roles in environmental management, agriculture, biomedical engineering, and textiles, and this appears to be driving the starch industry on the economic front.

As our understanding of starch and its applications continue to evolve, argument on its properties rages among biopolymer scientists; with every discipline defining starch from the prism or scope of such a discipline. For example, starch evolution has succeeded in moving it from its traditional role as food to being an indispensable medicine.

Starch - Evolution and Recent Advances is a unique collective project that brings together authors of varying backgrounds, experiences, and expertise in starch and starch products. It is organized into three sections: "Introduction," "History and Evolution," and "Recent Advances." Chapter 1 introduces the topic while also creating a nexus between starch and national development. The second and third sections contain five chapters each. Chapters 2–6 in Section 2 focus on the history, evolution, and ability of starch to serve as an economic powerhouse of nations. This section also expounds on the health benefits of starch as well as the effects of starch varieties and handling on the physicochemical properties of starch. Chapters also discuss the qualitative and quantitative morphological, chemical, and genetic properties peculiar to the source of the starch. Section 3 includes Chapters 7–11 and examines some of the most recent advances in starch and starch product technology, from the molecular behavior of starch and its effects on bioavailability, current and future applications of resistant starch as a functional food ingredient, and development of biodegradable starch films using electrospraying technology to the application of starch and starch-based polymeric materials in nanotechnology and environmental remediation.

This book provides the reader with excellent and rich knowledge of the history, evolution, and recent developments in starch. It is useful for students, pharmacists, agriculturists, chemists, biologists, chemical engineers, and health research scientists. University professors, research professors, researchers, and their undergraduate and postgraduate students will find this multi-authored, multi-sectoral, and multidisciplinary book useful when preparing their theses on starch and starch-related projects.

I wish to express my gratitude to the authors for their excellent contributions and to my very committed and understanding Author Service Manager Ms. Dolores Kuzelj at IntechOpen. Finally, I wish to thank my wife Comfort who endured many late nights of my editing this book.

Martins Ochubiojo Emeje
National Institute for Pharmaceutical Research and Development,
Abuja, Nigeria

Section 1

Introduction

Chapter 1

Introductory Chapter: What We Know about Starch

Martins Ochubiojo Emeje

1. Introduction

Several studies abound in literature expounding the meaning and frontiers of starch, which is the major dietary source of carbohydrates, and the most abundant storage polysaccharide in plants, occurring as granules in the chloroplast of green leaves and the amyloplast of seeds, pulses, and tubers. Starches are polysaccharides, made up of a number of monosaccharides or sugar molecules connected with a-D-(1–4) and/or a-D-(1–6) linkages. It consists of two major structural components: amylose, which is a linear polymer with glucose residues of α-(1–4) linked typically constituting 15–20% of starch; and the major component called amylopectin, which is a larger branched molecule with α-(1–4) and β-(1–6) linkages. The former is said to be linear or slightly branched. It is reported to have up to 6000 and 105–106 g/mol as its degree of polymerization and molecular mass respectively. The respective corresponding values for amylopectin are 2 million and 107–109 g/mol.; it is thus considered one of the largest naturally occurring molecules [1].

2. Starch and nations development

In the past two decades, the global market value for both native and modified starches has been put at 48.5 million tons worth €15 billion per year. Since the first century, Celsius, a Greek physician, had described starch as a healthy food, having been added to several foods, confectioneries, and wine as well as salad dressing ingredients in mayonnaise during the late 1800s in Germany and England. Later, combinations of cornstarch and tapioca were used by salad manufacturers. Starch is also used as a sweetener; Sweeteners resulting from the hydrolysis of starch with acid were used to improve wine in Germany in the 1830s.

Between 1940 and 1995, an increase in the use of starch in the US food industry was reported from about 30,000 to 950,000 metric tons. The main consumers of starch are believed to be the fermentation, leavening, and confectionary industries. A similar study in Europe in 1992 found that 2.8 million tons of starch were used in food. The uses of starch abound in literature, and the reader is advised to refer to reviews on the application of starch in the food industry. In fact, starch's versatility is unmatched by other biomaterials, making this polymer stand tall in the quest for natural products with high potentials to contribute to national development.

Local manufacturing is continually expanding, and there is heavy dependence on imported pharmaceutical-grade starch as a raw material for use in the manufacturing of various medicines used in healthcare delivery. This high dependence on foreign sources for raw material makes cost of production of the medicines used in healthcare delivery very high. Import dependence is the farmer's enemy while also hurting the country, because it replaces local production and creates unemployment, especially in developing countries of the world. This should not be acceptable, because it is neither economical or sustainable. Every country of the world has a number of locally available sources of starch, which if developed will reduce or eliminate heavy dependence on importation, thereby conserving foreign reserves, and invariably affects the cost of production of the medicines used in healthcare delivery. It is no brainer that local production of starch will create employment and equally put smiles on the faces of farmers. The production of pharmaceutical-grade starch will equally enable countries to compete in the global starch production [1–4].

3. Recent advances

In the last five decades, the general tendency in the starch industry has been in the utilization of raw materials other than maize and potato. Research has proven that the chemical composition and physical characteristics of a starch are essentially typical of its botanical and biological origin. It has also been found that, regardless of their origin, native starches are undesirable for many applications because they cannot withstand processing conditions such as extreme temperatures, variable pH, high shear rates, and frost variations. To overcome this challenge, changes are often made to improve or suppress the inherent properties of these native starches or to provide new properties that meet specific application requirements; such modified starches have found usefulness in the pharmaceutical, food, paper, and textile industries applications such as adhesives, disintegrants, fillers, emulsion stabilizers, and adhesives.

Although various methods of starch modification have been well documented in the literature, progress continues to be made in this area, and these changes have led to improved functional and physicochemical parameters of starch in various industries, especially where native starch itself cannot offer optimal performance. Recent studies have shown that starch is not just an important biopolymer because of its robust application in various spheres of life, but it has the potentials for use in sensitive and precision-driven areas such as in the preparation of various drug delivery systems; transitioning starch granules from its native micro- to the artificial submicron levels, thereby conferring on it new properties. Although carbohydrates are generally considered to be safe, their constituents especially at submicron level can present some safety issues, especially in pharmaceutical systems. It is only recently that the modified carbohydrate safety has been studied. This book provides detailed introduction of the historical and recent developments in carbohydrate research. It is clear that the demand for carbohydrates will continue to increase, especially with the use of this biopolymer in other industries. For example, in medical science, carbohydrates evolved from human food to effectively dry skin lesions. In the field of biological science, carbohydrates are also considered essential. Starch has today evolved from its traditional role as food to more modern roles in medicine, pharmacy, agriculture, life sciences, and the biopolymer industries [5–7].

4. Conclusions

In many countries of the world, especially developing countries, local drug manufacture is characterized by a complete dependence on importation of all pharmaceutical actives and excipients required for the manufacture of drugs and cosmetics, and starch is key among these excipients. Many of these countries have the challenge of not only the cost of foreign exchange in addition to the lack of infrastructure, which not only results in a very high cost of production but also leads to low capacity utilization. This situation can be rectified if the local raw materials available in these countries are processed into pharmaceutical-grade raw materials. This would have a direct impact on the national economy by enhancing natural resource utilization. The problem of locally producing pharmaceutical grade starch can be tackled in either of two ways, i.e. either undertake primary production of starch as required by the pharmaceutical industry or by improving available grades of commercial starch to officially acceptable pharmaceutical grade. It has been opined that the second option should be cheaper, therefore, undertaking studies of locally produced starches with a view to identifying areas where they fall short of pharmaceutical-grade and subsequently determine methods required to correct such that would lead to the upgrading of locally produced starch to pharmaceutical-grade starch is a viable strategy, and this book will serve as a veritable tool in the hands of students, researchers, and industry players.

Author details

Martins Ochubiojo Emeje
National Institute for Pharmaceutical Research and Development (NIPRD), Abuja, Nigeria

*Address all correspondence to: martinsemeje@yahoo.com

IntechOpen

© 2022 The Author(s). Licensee IntechOpen. This chapter is distributed under the terms of the Creative Commons Attribution License (http://creativecommons.org/licenses/by/3.0), which permits unrestricted use, distribution, and reproduction in any medium, provided the original work is properly cited. (cc) BY

References

[1] Schwartz D, Whistler RL. Chapter 1-History and future of starch. In: BeMiller J, Whistler R, editors. Food Science and Technology, Starch (Third Edition). London, United Kingdom: Academic Press; 2009. pp. 1-10. DOI: 10.1016/B978-0-12-746275-2.00001-X

[2] Fortunak JMD, Emeje MO, Kammendi H, Tilahun EL, Wang XS. Chapter 13. The business case for green chemistry in drug discovery. In: Green Chemistry Strategies for Drug Discovery. Piccadilly, London: The Royal Society of Chemistry; 2015. pp. 280-313

[3] Ochubiojo EM, Rodrigues A. In: Valdez B, editor. Starch: From Food to Medicine, Scientific, Health and Social Aspects of the Food Industry. Rijeka: InTech; 2012. pp. 355-380. Available from: http://www.intechopen.com/articles/show/title/starch-from-food-to-medicine

[4] Sweedman MC, Tizzotti MJ, Schäfer C, Gilbert RG. Structure and physicochemical properties of octenyl succinic anhydride modified starches: A review. Carbohydrate Polymers. 2013;**92**(1):905-920

[5] Rodrigues A, Emeje M. Recent applications of starch derivatives in nanodrug delivery. Carbohydrate Polymers. 2012;**87**:987-994. DOI: 10.1016/j.carbpol.2011.09.044

[6] Emeje MO, Obidike IC, Akpabio EI, Ofoefule SI. Nanotechnology in drug delivery. In: Sezer AD, editor. Recent Advances in Novel Drug Carrier Systems. Rijeka: InTech; 2012. pp. 69-106. Available from: http://www.intechopen.com/articles/show/title/nanotechnology-in-drug-delivery

[7] Onyango C. Starch and modified starch in bread making: A review. African Journal of Food Science. 2016;**10**(12):344-351. DOI: 10.5897/AJFS2016.1481

Section 2

History and Evolution

Chapter 2

History, Evolution and Future of Starch Industry in Nigeria

Obi Peter Adigwe, Judith Eloyi John and Martins Ochubiojo Emeje

Abstract

Starch industry has progressed into a business that is worth billions of dollars globally, as they have been found useful in the food, textile, biofuel, plastic and the pharmaceutical industries. Nigeria can be the largest producer of starch in the world. Her major sources are roots and tubers (cassava, yam, cocoyam and potato), cereals (maize, sorghum, millet and rice) and fruits (banana, plantain and breadfruit). Although, all the starch crops are abundantly produced in Nigeria, only less than 1% is processed into high quality starch for industrial processes. This chapter therefore examines the past, the progression and the current state of the starch industry in Nigeria and the roles the government and relevant stakeholders must play in order to revolutionize the industry in Nigeria.

Keywords: starch, cassava starch, rice starch, potato starch, maize starch, industrial uses, history, evolution, Nigeria

1. Introduction

Starch is the most abundant edible polysaccharide derived from plants [1]. They are composed of repeating units of amylose and amylopectin that are susceptible to modification by physical, chemical, enzymatic and other means [2]. Starch in its native form is a multipurpose polymer, and an important raw material that finds useful applications in both food and non-food processes [1, 3]. Native starch is susceptible to retrogradation, syneresis, inconsistent viscosity and unordered gelation, hence, cannot withstand the typical industrial processes such as high temperatures, pH, high shear rate, and freeze thaw variations usually encountered during its use in food and other processes. Thus modification transforms native starch into gels or viscous mass in water at room temperature [2]. Furthermore, their properties are improved to yield starches with enhanced pasting properties, decreased retrogradation, reduced tendency to gel, enhanced freeze-thaw stability, improved paste clarity, and texture, and improved film formation and adhesion. These superior properties are what finds application in the food, textile, cosmetics, agro and pharmaceutical industries. The major sources of starch are roots and tubers (cassava, yam, cocoyam and potato), cereals (maize, sorghum, millet and rice) and fruits (banana, plantain and breadfruit) [4, 5]. However, commercially available starches are cassava, maize, potato, and rice starch and they are gluten-free which can be tolerated by individuals who react to gluten [6]. Even though all the starch crops are abundantly produced in Nigeria, only less than 1% is processed into high quality starch for industrial processes, while about 10% is used as feedstock, 5% is processed into syrup concentrates for the soft drink industry and the rest (84%) is consumed as food [2].

1.1 Cassava starch

Cassava (*Manihot esculenta*), is extensively cultivated as an annual crop in tropical and subtropical regions, around the South-East, South-South and South-West regions (**Figure 1**), and it serves as an important food crop and a major source of carbohydrate for more than 70% of the Nigerian population, and provide income to over 30 million farmers, industries and traders [7, 8]. Nigeria is the largest producer of cassava in the world, producing over 37 million metric tons with an average yield of 12 metric tons/hectare [9]. Over 90% of these are consumed locally, and used as animal feeds, industrial purposes in pharmaceutical and food industry and only 10% are processed for export [3, 9]. Cassava root contains about 1% protein and 30–35% starch on a dry weight basis, hence, regarded as a starchy food [10]. It is a perishable commodity once harvested with a shelf life of less than 3 days [7]. The greatest potential of cassava as an agricultural crop lies in the production of starch. Two types of starch (sweet and sour) are produced from both varieties of cassava tubers (bitter or sweet). Both forms of starch differ in characteristics and end use. Sweet starch is obtained after an extraction process that separates the starch from other constituents, whereas sour starch is obtained by fermentation following extraction. Sweet starch is used as an adhesive in the textile, paper, and battery industries, while the sour starch is used solely in the food industry [3]. Although cassava starch serves as a cost effective raw material for many industrial processes [1], studies show that 52% of cassava produced is wasted as a result of inadequate production and processing, 43% is consumed as food and 5% is utilized as feedstock [3]. The suitability of cassava starch as a raw material in the food industry is due to its inherent properties such as low gelatinization temperature (71°C), its slow propensity to retrograde, does not have leftover proteinous materials or soil residues, non-cereal flavor, high viscosity, high water binding capacity, bland taste, translucent paste, good stability and reasonably good adhesive strength. It has a rather low lipid profile of less than 0.01%, protein content of 0.15–0.30%, ash value of 0.08–0.15% and phosphorous content of 2.04–2.45 mg/kg. In addition, it is easy to extract, settles rapidly and has a carbohydrate content of 73.7–84.9% [11]. In the paper industry, cassava starch is used as an adhesive; in the cosmetic industry, it serves as a significant raw material in the production of powders, it enhances the recovery and stability of detergents in the soap industry, while it produces a better foaming ability in the rubber and foam industries. In addition, cassava starch can be converted into maltotriose, maltose, and glucose, can be used to make fructose syrups and in the production of gelatin capsules [1]. Although, cassava starch has great potentials, only about 2% of food industries in Nigeria utilize cassava starch and its derivatives [3].

1.2 Potato starch

Potato (*Solanum tuberosum L.*) belongs to the tuber crops. There are two major types: the Irish potato which are grown through tubers, and the sweet potato (*Ipoema batata*), raised through vines, along the North-Western part of Nigeria (**Figure 1**). The Irish potato was introduced first in Nigeria in the late 19th Century, through missionary activities. Encouraged by the British government during the Second

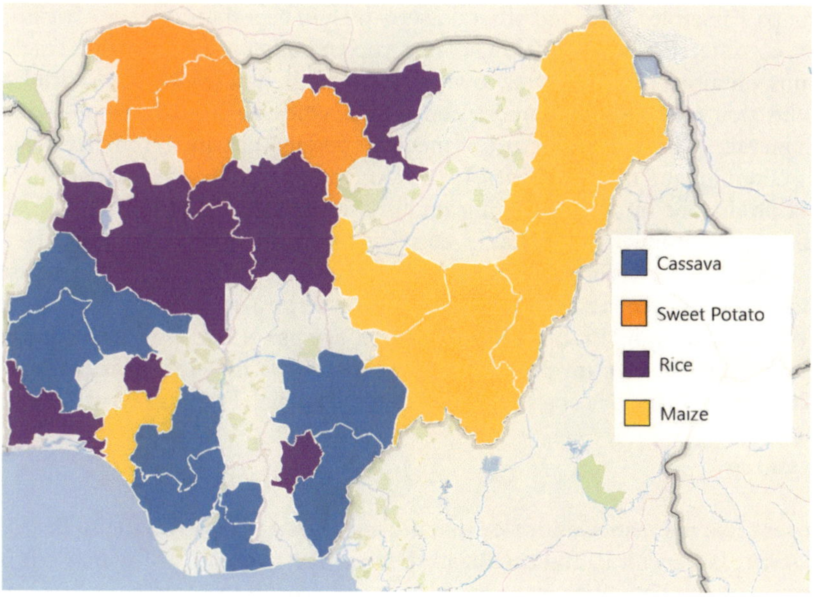

Figure 1.
Map showing regions of some starch-based crops in Nigeria.

World War in order to feed their troops in West Africa, Irish potato was cultivated extensively and it became an important commodity for both local and international trade [12]. Nigeria is known to be the fourth country in West Africa in terms of potato production; on the African continent, it ranks the seventh. However, little effort is made in its promotion for industrial use, as over 80% are consumed as food [13]. Starch obtained from Potato has been demonstrated as a potential raw material in the pharmaceutical, textile, wood and paper industries as an adhesive, binder, texture agent and filler, and by oil drilling firms to wash boreholes [12]. Sweet potato starch can be utilized in the production of starch syrups, glucose and isomerized glucose syrups, lactic acid beverages, bread and other confectionaries [5, 14]. It can also be used to produce distilled spirits, noodles and isomerized saccharides as sweetener for soft drinks [15]. In addition, they can be utilized in the production of citric acid, ethanol and also used in the paper and textile industries [16]; as stabilizer in the production of yogurt [17]; and as pharmaceutical excipients when modified [18–20]. Potato starch is 100% biodegradable and is utilized as a substitute for polystyrene and other plastics. Hence, can be used in the production of disposable plates, dishes and knives [12].

1.3 Maize starch

Maize grains (*Zea mays*) are widely distributed worldwide, and grown in abundance in the North-Western and Southern-Western parts of Nigeria (**Figure 1**) during rainy seasons, however, they are subject to post-harvest wastes due to inadequate storage and processing. More maize is produced annually than any other grain. About fifty (50) species exist and consist of different colors, textures and grain shapes and sizes. White, yellow and red are the most predominant types. Maize starch or corn starch is the starch derived from the corn (maize) grain. The starch is obtained from the endosperm of the corn kernel and is a popular food ingredient used in the food, textile, pharmaceutical and paper industries. In the food industry, is used as a thickening agent for sauces, gravies, glazes, soups, casseroles, pies, and other desserts [21]. In Nigeria and other African countries, corn starch is used in making corn syrup and other sugars like high-fructose corn syrups, obtained from the breakdown of corn starch, utilized in the soft drink and candy industries. And also in the production of bioplastics. It is equally the preferred anti-stick polymer used in the manufacture of medical products obtained from natural latex, including condoms and medical gloves. In the food industry, corn starch is used to reduce the cost of production by adding varying amounts of corn starch to foods like cheese and yogurt. In the production of ethanol, corn has the least expensive total cost. Here, the yellow specie of corn is used as it contains about 62% of starch [22]. In the industrial production of glucose, corn starch undergoes hydrolysis by a degradation process using amylolytic enzymes found in abundance in nature. The use of enzymes in this process produces a higher yield of pure glucose, that are more stable and environmentally friendly [23]. In 2004 more than 50% of starch was converted to High Fructose Syrups (HFS).

1.4 Rice starch

There are two main species of rice that are cultivated in Nigeria, the African specie, *Oryza glaberrima* L. and the Asian specie, *Oryza sativa* L., of which 120,000 varieties are known [24]. Rice is the world's second most important cereal crop after maize, based on the volume of production and cultivated mainly along the North-West and South-West regions of Nigeria. Nigeria currently consumes about 7.9 million metric tons of rice annually while production is presently at 5.8 million metric tons. The FAO's report show that rice generates more income for Nigerian farmers than any other cash crop in the country, with small scale farmers accounting for the largest volume in sales of 80% while the remaining 20% is consumed. An average of 7–8 tons of rice can be obtained per hectare. But up to 12 tons per hectare can be obtained depending on the variety of seeds planted. Although rice has been the topic for discussions, it has not benefited from the kind of value-added research required for economic competitiveness on an international scale, in terms of production. Hence, rice producers in the country are peasant farmers who are left to keep the sub-sector afloat against all odds. Industrialists can be encouraged to engage in large-scale production of rice as functional ingredients can be developed from Nigeria local rice cultivars and that these ingredients would stand the world market competitiveness. Of these, rice is starch is the major component of rice constituting about 90% of its dry matter [25]. Another major product obtained from rice starch is liquor, usually called Rice Wine, and it can be made at home or in a processing facility from the fermentation of rice starch that has been converted to sugar. It is widely consumed in Asia, and has an average alcohol content of 18–25%. At the present, this area of investment is yet to be explored in Nigeria [26]. Traditionally, there have been basic attributes associated with rice starch that makes it stand out above other cereal and non-cereal starches. These properties include hypoallergenicity, digestibility, bland flavor, small granule (3–10 μm), white color, greater acid resistance, greater freeze–thaw stability of pastes and a wide range of amylose/amylopectin ratios. These exceptional features are manifested in the different applications of rice starches.

2. History of starch industry in Nigeria

In Nigeria, in 1940, starch began gaining popularity when cassava starch was produced in response to the demand by the British government during an outbreak of war. In May 1940, starch samples from cassava roots were sent to the Ministry of Food and to Starch Products Limited in London for further experiments [27], and 300 tons of starch was purchased. Following the successful sale of starch and its increased demand by the British government, 10,000 tons of starch was further exported in July, August, and September 1940. However, in October, 1940, there was a drawback in the export of starch due to lack of quality, as a result of delay in processing cassava roots, inadequate washing of the starch after settling, the use of dirty water or dirty utensils, the use of exposed peeled roots to the atmosphere for undue time before grating, storage of starch with high moisture content [27].

By June, 1941, the Ministry of food in London purchased starch "irrespective of quality"; and by September the UK was buying the entire production obtainable [28]. Trade continued in 1942, with the demand of monthly supply set at 300 tons by the Ministry of Food, this continued into 1943, and from January to April, starch was produced in large quantities and became a priceless commodity [27]. In April, 1943, exportation of starch from Nigeria was abolished as a result of a new demand for palm products by the UK. Therefore, in 1943, Nigeria was required to ensure the maximum production of palm products, and to achieve this sole objective, a strategy was implemented to stop the production of starch [29]. Likewise, the oil and kernel producing areas in Nigeria were barred from producing starch, instead, they were encouraged to cultivate cassava enough for local consumption [27, 30]. In May, 1943, the purchase of starch for export ceased and the notice on prohibition was repeated in 1944 and 1945 [27, 31]. Since then, low quality starch was refined and sold locally to be consumed as food or adulterated with cassava flour or garri and sold in the markets [32].

3. Evolution of starch industry in Nigeria

Starch industrial application has evolved into a multibillion dollar business worldwide and as such, many more industries, mostly within Africa, have now developed multipurpose applications for starch especially cassava starch. The demand for starch in Nigeria alone has recently been estimated to be around 67,100 tons per year and the amount of fresh cassava roots needed to produce that amount of starch is 350,000 tones [1]. The production of cassava has been stagnant since independence to the mid-1980s. however, an increase in production was observed from the mid-1980s to early 1990s where it remained constant again until 1999, at the beginning of the civilian era, where an increase was observed. Although, the Nigerian cassava market is centered on consumption patterns, an industrial market is now evolving (where cassava starch is used in the food, pharmaceutical and other industries) that needs to be explored and utilized. Starch derived from cassava for industrial purposes reveals a great potential for increased earnings for cassava farmers. Howbeit, in order to achieve global competitiveness on the production of cassava, the Obasanjo administration in 2002 established an ordinance on cassava production that led to a 73% growth, an excess in cassava production, which eventually led to wastage and massive financial loss due to poor storage, inadequate distribution and underdeveloped downstream sector [33, 34]. This led to investments in new factories (MATNA foods in Ondo state, built in 2005, Dutch Agricultural Development Company Nigeria Limited, Benue State, built in 2006, and Ekha Agro Farms, Ogun State, a glucose syrup factory built in 2007). These companies used improved production and processing machineries to enhance cassava starch production and also in the production of glucose, sweeteners and ethanol that could meet international standards [33]. Likewise, the annual production of rice in Nigeria increased from 2.8 million in 2010 to 4.9 million metric tons in 2019, while rice importation decreased by about 60% within the time frame, indicating Nigeria's readiness to achieve self-sufficiency. The improved development warranted product diversification of rice to other value-added products like ethanol, glucose syrup, and starch by making use of the underutilized native rice varieties that had undesirable physical features (short grain length, poor color, etc.), poor cooking quality (soft and sticky grain), and poor consumer acceptability [35].

4. Prospects of starch industry in Nigeria

Native starch is undesirable for many industrial applications, irrespective of the source. They are susceptible to retrogradation, syneresis, undesirable viscosity and gelation as a result of their inability to withstand high temperatures, pH, high shear rates, and freeze thaw variations when used for food and other applications. Hence, the need for modification by physical, chemical or biological means to yield starch with improved pasting properties, decreased retrogradation, decreased tendency to gel, increased paste freeze-thaw stability, improved paste clarity, film formation, adhesion and gel texture [36].

4.1 The food industry

The food industry is one major industry that utilize starch and starch-based products. Modified starches have been used as sweeteners, and to improve the texture of gums and pastes, and also to obtain products that thicken in cold water without the addition of heat [2]. They are also used as binders to solidify the mass of food to prevent it from drying out during cooking especially in sausages and processed meats; and as a stabilizer in creams, due to its high water-holding capacity [37, 38]. As thickeners, they are used in soups, baby foods, sauces and gravies. In addition, glucose produced from starch are also used in the production of caramel that are extensively utilized as coloring agents in foods, confectionaries and liquor [39]. Glucose syrup is a solution of glucose, maltose and other nutritive saccharides obtained from edible starch. Although, glucose syrup is not adequately produced in Nigeria as 800 million naira was used to import glucose in 2003 [4]. However, a surge in private sector investments in the large-scale production of starch as reported by the Federal Ministry of Agriculture, Nigeria was observed and companies like the Nigeria starch mills, and the glucose factory in Ogun State began to invest in large-scale production of cassava and processing of its by-products by utilizing improved techniques in the machineries used to peel, grate, dry, fry and mill the raw materials [40].

4.2 The pharmaceutical industry

In the pharmaceutical industry, starches are one of the most important excipients that have been widely used in the formulation of tablets. They are inert, inexpensive and have been utilized as fillers, binders, disintegrants and glidants. The use of modified starch improves the physicochemical properties of the tablets and other pharmaceutical formulations [36, 41]. Over 1500 tons of starch hydrolysates are used in the pharmaceutical industry annually and about 80% of starch are imported [8]. In addition, crystalline and liquid glucose are imported for the production of cough syrups, yet, this high demand for starch-based products can be met from the inexpensive and high quality starch found in cassava that is richly available in Nigeria [1].

4.3 The textile industry

In the textile industry, modified starch (oxidized starch) is employed in the sizing and dyeing of fabrics to improve the weight, clarity and hardness of the fabric [1, 36]. Cassava starch is the most preferred for this application as it gives a better finishing compared to other starches. In addition, modern laundries use soluble starch, incorporated into a suitable propellant sprays for application during steam ironing. In the early 1990_s only about 700 tons of cassava starch was produced per annum because the starch was considered to be of low quality. However, maize starch was imported for use in the textile industry which was later replaced with over 67,000 tons of cassava starch [8]. Again, the downturn of the economy during the military era led to the near collapse of the textile industry which further decreased the market prospect for cassava starch. Nevertheless, this industry can be restored if successive leadership can establish and promote the starch industry [42].

4.4 The plastic industry

Starch has been used in the production of plastics since 1970s based on their biodegradability, renewable nature and freely abundant. Starch is easily dispersed in cold water, however, it thickens when heated to its near boiling point to form a colloidal suspension that gels when cooled. Polymer blends in the presence of plasticizers like water and glycerol, can be distributed or transported to normal plastic converters, which can process the blends to products using normal injection or blow molds [43]. Such biopolymers (consisting of 40% starch and 60% low-density polyethylene) using cassava starch has been produced and commercialized [1]. There are roughly 30 plastic companies in Nigeria, among which are: The Black Horse Plastic Industry Limited, OK Plast Limited, Abbey-Fem Plastics, Celplas Industries Nigeria Limited, among others. Since Nigeria is the largest producer of cassava, the utilization of cassava starch by these companies to produce biodegradable biopolymers is achievable. This will not only protect the environment from the harmful effects of petroleum derived plastics, but will also generate an economic alternative for cassava agriculture [44, 45].

4.5 The biofuel/chemical industry

In the production of alcohol, starch is hydrolyzed by a two-step process to glucose and then is further diluted and converted to ethanol by the action of yeast [3]. Bioethanol is a form of soi-disant-renewable energy that can be manufactured from agricultural crops like corn, potatoes, cassava, rice and sugar cane [46, 47]. Cassava starch has a much higher yield (150 L/ton of fresh roots) than sugar cane (48 L/ton), a source that was previously used in the production of ethanol without much success in Nigeria. However, difficulty in the local production of cassava in 2001 halted the production of ethanol and since then, all the ethanol used in Nigeria was imported. Although current interests in investment in the Nigerian ethanol industry is increasing, encouraging small industrial production using cassava will lead to the economic growth of the industry [8]. Other starch-derived products such as D-glucose and maltose, butanol, acetone, glycerol acetic, citric, itaconic, gluconic and lactic acids can also be produced from these starch-based crops by the process of fermentation when the starch is modified [2]. According to **Amenaghawon et al.** [48], the projected yearly production of citric acid is about 1.4–1.5 million tons and this is expected to increase to about 3.5–4.0% yearly. Of this amount, roughly 70% is used by the food industry for its pleasant acid taste, high solubility in water, its chelating, antioxidant and buffering actions. Approximately 12% is used by pharmaceutical industries as liquid elixirs, flavoring agents, anti-coagulant and as preservatives while 18% is used by other industries (cosmetics, toiletry, detergent, textile, oil recovery, paper).

4.6 Other starch-based products

Other uses of starch are in the production of adhesives; starch (converted to dextrin) is the major raw material in the manufacture of glues and adhesives [1, 42]. In the 1990s, 58,000 tons of adhesives were used in the wood, cable, paper and printing, packaging and footwear industries in Nigeria. Regrettably, they were imported either as adhesives or as dextrin. Adekunle et al. [49–52] demonstrated the potentials of using cassava starch as a raw material in the large-scale manufacture of adhesives in Nigeria. Hence, expanding the starch industry for use in the manufacture of adhesives for these industries would put over 60,000 tons of cassava into use for this industry alone in Nigeria [8]. In the soap and detergent industries, starch is used as a filler to obtain a better yield and enhance the shelf-life of the products; in the production of sugar syrups, starch is subjected to enzyme hydrolysis using α-amylase. Cassava, corn and rice starch have been used in the production of fructose and glucose syrups, and in the manufacture of gelatin capsules [42, 53]. The food and beverage industries in Nigeria depend heavily on glucose syrups and crystalline sugars, and cassava starch are used in the production of candies, in the soft drink industries and in traditional medicines. However, the syrup concentrates are currently imported as cassava starch derivatives (hydrolysates e.g. glucose, sucrose, fructose, maltose, and syrup) are not presently developed in Nigeria [1, 8]. In the production of yeast, starch is enzymatically hydrolyzed to glucose and this leads to the production of certain yeasts that utilize this glucose to produce microbial cellular substances. According to Taiwo et al. [3], this aspect of producing yeast from simple sugars is yet to be exploited in Nigeria, as majority of the yeast used in the food and beverage industries are imported.

5. The future of starch industry in Nigeria

To achieve sustainable progress in the starch industry in Nigeria, the government, private sectors and key stakeholders must of necessity put all hands on deck to ensure the improvement of the agricultural sector in Nigeria [54]. This can be achieved by implementing programs and policies that will foster the production of starch producing crops, conversion of unused lands into cultivation of starch-based crops, establishment of industries for local production of starch, and improvement in the processing and storage of starch-based crops [1]. Thus, the local industrial products will not only meet local demands but also have the potential of becoming a source of income generation. This demand will aid in the improvement of industry, national and international standards, quality and global supplies leading to a gradual approach to export-oriented production. In addition, private investors may be encouraged to participate in improving the production and conversion of these crops to high quality starch and derivatives, by supporting market linkages, organizing trainings for farmers, support and development of processor groups, build capacity in quality and standards in product development and intellectual property rights [55]. Government on the other hand can strengthen the policy on the development of starch and starch-based products from the available sources that can be effectively used in the food, feed, textile and pharmaceutical industries [56]. Infrastructure especially electricity supply, railway transport, and water supply can also be improved to achieve success [57].

6. Conclusion

Nigeria is endowed with starch crops, and ranks third after wood and vegetable oil. it is highly beneficial and an invaluable commodity to the pharmaceutical, textile, food, paper, adhesive, drinks, beverages and the confectionery industries. It can compete favorably on international scale, however, starch in Nigeria is underutilized. The cassava starch is preferred over the corn starch in Nigeria as it is the driving force for the conservation of foreign exchange and to reduce import dependency. The opportunities in the starch industry is incredible and the demand for starch is on the rise, this can create income-generating opportunities for youths and small-scale farmers in the country. Hence, efforts should be made by the government and relevant stakeholders to convert starch-based crops from low-yielding famine crops to high-yielding cash crops in order to foster the economic development of the country.

Conflict of interest

The authors declare no conflict of interest.

Author details

Obi Peter Adigwe, Judith Eloyi John* and Martins Ochubiojo Emeje
National Institute for Pharmaceutical Research and Development (NIPRD), Abuja, Nigeria

*Address all correspondence to: juditheloyi@gmail.com

IntechOpen

© 2022 The Author(s). Licensee IntechOpen. This chapter is distributed under the terms of the Creative Commons Attribution License (http://creativecommons.org/licenses/by/3.0), which permits unrestricted use, distribution, and reproduction in any medium, provided the original work is properly cited.

References

[1] Tonukari NJ. Cassava and the future of starch. Electronic Journal of Biotechnology. 2004;7(1):1982-1985

[2] Daramola B, Falade KO. Enhancement of agronomical values: Upstream and downstream opportunities for starch and starch adjuncts. African Journal of Biotechnology. 2006;**5**(25):2488-2494

[3] Taiwo KA. Utilization potentials of cassava in Nigeria: The domestic and industrial products utilization potentials of cassava in Nigeria. Food Review International. 2006;**22**:29-42

[4] Elizabeth Ojewumi M, Adewale Adeeyo O, Mary Akingbade O, Elizabeth Babatunde D, Ayodele Ayoola A, Olufemi Awolu O, et al. Evaluation of glucose syrup produced from cassava hydrolyzed with malted grains (rice, sorghum & maize). International Journal of Pharmaceutical Sciences and Research. 2018;**9**(8):1000. DOI: 10.13040/IJPSR.0975-8232.9

[5] Tunde AA. Production of glucose from hydrolysis of potato starch. World Scientific News. 2020;**145**(April):128-143

[6] Mohamad Yazid NS, Abdullah N, Muhammad N, Matias-Peralta HM. Application of starch and starch-based products in food industry. Journal of Science and Technology. 2018;**10**(2):144-174

[7] Dada AD. Taking local industry to global market: The case for Nigerian cassava processing companies. Journal of Economics and Sustainable Development. 2016;**7**(19):59-70

[8] Ezedinma CI, Kormawa PM, Manyong VM, Dixon AGO. Challenges, opportunities, and strategy for cassava sub sector development in Nigeria. In: Proceedings of the 13th ISTRC Symposium. Arusha, Tanzania: International Society for Tropical Root Crops; 2007. pp. 627-640

[9] Awerije BO. Technical, cost and allocative efficiency of processing cassava into gari in Delta State, Nigeria. 2016. pp. 1-24

[10] Ogunmuyiwa OH, Adebowale AA, Sobukola OP, Onabanjo OO, Obadina AO, Adegunwa MO, et al. Production and quality evaluation of extruded snack from blends of bambara groundnut flour, cassava starch, and corn bran flour. Journal of Food Processing and Preservation. 2017;41(5):1-11

[11] Falade KO, Akingbala JO. Utilization of cassava for food. Food Review International. 2011;**27**(1):51-83

[12] Ugonna C, Jolaoso MO, Onwualu AP. A technical appraisal of potato value chain in Nigeria. International Research Journal of Agricultural Science and Soil Science. 2015;**3**(8):291-301

[13] Tewe OO, Ojeniy FE, Abu OA. Sweet Potato Production, Utilization, and Marketing in Nigeria. Lima, Peru: Social Sciences Department, International Potato Center (CIF); 2003. pp. 1-52

[14] Etudaiye HA, Oti E, Aniedu C. Utilization of sweet potato starches and flours as composites with wheat flours in the preparation of confectioneries. African Journal of Biotechnology. 2015;**14**(1):17-22

[15] Odebode SO, Egeonu N, Akaroda MO. Promotion of sweet potato for food industry in Nigeria. Bulgarian

Journal of Agricultural Science. 2008; **14**(3):300-308

[16] Betiku E, Adesina OA. Optimization of sweet potato starch hydrolyzate production and its potential utilization as substrate for citric acid production. British Biotechnology Journal. 2013; **3**(2):169-182

[17] Altemimi AB. Extraction and optimization of potato starch and its application as a stabilizer in yogurt manufacturing. Foods. 2018;7(14):1-11

[18] Mohammed KG. Modified starch and its potentials as excipient in pharmaceutical formulations-mini review related papers modified starch and its potentials as excipient in pharmaceutical formulations. Novel Approaches in Drug Designing & Development. 2017;**1**(1):1-5

[19] Jubril I, Muazu J, Mohammed GT. Effects of phosphate modified and pregelatinized sweet potato starches on disintegrant property of paracetamol tablet formulations. Journal of Applied Pharmaceutical Science. 2012;**2**(2):32-36

[20] Odeku OA. Potentials of tropical starches as pharmaceutical excipients: A review. Starch/Staerke. 2013;**65**(1-2): 89-106

[21] Fora/Cornstarch/FMR/165390. Corn (Maize) Starch in Nigeria. The Feasibility Report. 2016

[22] Clifford CB. Alternative Fuels from Biomass Sources. 2017.

[23] Ubalua AO. The use of corn starch for growth and production of α-amylase from *Bacillus subtilis*. Journal of Microbiology Research. 2016;**4**(4):153-160

[24] Ombretta M, Valeria S, Dayana C, Giuseppe P. The use of rice in brewing. In: Advances in International Rice Research. UK: Intech Open; 2017. pp. 49-66. DOI: 10.5772/66450. Chapter 4 Abstract

[25] Akintayo E, Ashogbon A. Morphological, functional and pasting properties of starches separated from rice cultivars grown in Nigeria. International Food Research Journal. 2012;**19**(2): 665-671

[26] Nextzon. Rice Production in Nigeria. 2017. pp. 1-12

[27] Falola T. Cassava starch for export in Nigeria during the second World War. African Economic History. 1989;**18**(18): 73-98

[28] David K, Richard R. Africa and the Second World War. London: Palgrave Macmillan; 1986. p. 1-294.

[29] Falcon WP, Jones WO, Pearson SR, Dixon JA, Nelson GC, Roche FC, et al. The Cassava Economy of Java. Stanford: Stanford University Press; 1984. pp. 1-225

[30] Kilby P. The Nigerian Palm Oil Industry. 1967

[31] Byfield JA. Feeding the troops: Abelokuta (Nigeria) and World War II. African Economic History. 2007;**35**(35): 77-87

[32] Falola T. "Salt is Gold": The management of salt scarcity in Nigeria during World War II. Canadian Journal of African Studies. 1992;**26**(3):412-436

[33] Saweda L, Liverpool O. Enhancing the Competitiveness of Agricultural Commodity Chains in Nigeria: Identifying Opportunities with Cassava, Rice, and Maize using a Policy Analysis Matrix (PAM) Framework. 2009. pp. 1-54

[34] Mohammed AI. The dependency syndrome and Obasanjo's national cassava policy in Nigeria. International Journal of Advanced Academic Research. 2018;**4**(11):1-16

[35] Ofoedu CE, Osuji CM, Omeire GC, Ojukwu M, Okpala COR, Korzeniowska M. Functional properties of syrup from malted and unmalted rice of different varieties: A comparative study. Journal of Food Science. 2020; **85**(10):3081-3093

[36] Olu-owolabi BI, Afolabi TA, Adebowale KO. Effect of heat moisture treatment on the functional and tabletting properties of corn starch. African Journal of Pharmacy and Pharmacology. 2010;**4**(7):498-510

[37] Okechukwu PE. Influence of granule size on viscosity of cornstarch suspension. Journal of Texture Studies. 1995;**26**:501-516

[38] Omoregie EH. Chemical properties of starch and its application in the food industry. In: Chemical Properties of Starch. UK: Intech Open; 2020. pp. 1-26

[39] Akusu OM, Emelike NJT. Fermentation of corn starch powder for the production of "Ogi". Journal of Food Research. 2018;**7**(5):49-56

[40] Sanni LO. Recent Developments in Cassava Processing, Utilization and Marketing in Nigeria and Lessons Learned. 2008. pp. 1-11

[41] Adetunji OA, Odeniyi MA, Itiola OA. Compression, mechanical and release properties of chloroquine phosphate tablets containing corn and trifoliate yam starches as binders. Tropical Journal of Pharmaceutical Research. 2006;**5**(2): 589-596

[42] Tonukari NJ, Tonukari NJ, Ezedom T, Enuma CC, Sakpa SO, Avwioroko OJ, et al. White gold: Cassava as an industrial base. American Journal of Plant Sciences. 2015;**06**(07):972-979

[43] Duduyemi O, Mojibayo I, Shonaike G, Olodu OE, Babatunde BH, Adaran AS, et al. Effects of copolymer blends in the production and characterisation of biodegradable polymer from agricultural product-using cassava starch (Manihort Species) as case studies. Greener Journal of Agricultural Sciences. 2020;**10**(1):51-56

[44] Abioye OP, Abioye AA, Afolalu SA, Ongbali SO. A review of biodegradable plastics in Nigeria. International Journal of mechanical Engineering and Technology. 2018;**9**(10):1172-1185

[45] Eterigho EJ. Production and characterization of biodegradable plastic from Nigeria cassava starch. In: Proceedings of the International Conference on Engineering, Science, and Applications. Tokyo, Japan: Global Academic-Industrial Cooperation Society (GAICS); 2018. pp. 78-85

[46] Mbonu OF, Udeozor PA, Umoru GU, Uti DE. Production of Bio-fuel from sweet corn (food to fuel). Journal of Pharmacognosy and Phytochemistry. 2016;**5**(6):43-47

[47] Nwaokocha CN, Giwa SO, Layeni AT, Kuye SI, Oyedepo SO, Lawal NS, et al. Energy generation from corn starch effluent using microbial fuel cell using lead electrodes. Advances in Electrical and Electronic Engineering. 2021;**1**:1-11

[48] Amenaghawon NA, Aisien FA. Modelling and simulation of citric acid production from corn starch modelling and simulation of citric acid production from corn starch hydrolysate using

Aspergillus niger. Environment and Natural Resources Research. 2012; **2**(1):1-14

[49] Adekunle OD, Kayode OM, Awoyale OK, Dawodu M. Extraction, characterization and dextrinization of starch from six (6) varieties of tubers from Iwo Osun State Nigeria for application in the production of adhesives. American Journal of Chemical and Biochemical Engineering. 2019;**3**(2):7-11

[50] Akpa J. Production of starch-based adhesives. Research Journal in Engineering and Applied Sciences. 2012;**1**(4):219-214

[51] Chukwuemeka IS, Ugochukwu IW. Production of adhesive from cassava starch in Owerri, Imo State, Nigeria. World News of Natural Sciences. 2017;**11**:5-10

[52] Onyenwoke CA, Simonyan KJ. Cassava post-harvest processing and storage in Nigeria: A review. African Journal of Agricultural Research. 2014;**9**(53):3853-3863

[53] Zainab A, Modu S, Falmata AS. Laboratory scale production of glucose syrup by the enzymatic hydrolysis of starch made from maize, millet and sorghum. Biokemistri. 2011;**23**(1):1-8

[54] Daneji MI. Agricultural development intervention programmes in Nigeria (1960 to date): A review. Savannah Journal of Agriculture. 2011;**6**(1):101-107

[55] Olaoye OA. Potentials of the agro industry towards achieving food security in Nigeria and other Sub-Saharan African countries. Journal of Food Security. 2014;**2**(1):33-41

[56] Nchuchuwe FF, Adejuwon KD. The challenges of agriculture and rural development in Africa: The case of Nigeria. nternational Journal of Academic Research in Progressive Education and Development . 2012; **1**(3):45-61

[57] Adenle AA, Manning L, Azadi H. Agribusiness innovation: A pathway to sustainable economic growth in Africa. Trends in Food Science and Technology. 2017;**59**:88-104

Chapter 3

Starch: A Veritable Natural Polymer for Economic Revolution

Obi Peter Adigwe, Henry O. Egharevba and Martins Ochubiojo Emeje

Abstract

Amidst growing concerns for environmental degradation by anthropologic activities and use of non-biodegradable materials for industrial and household purposes, a focus on natural polymeric materials offers the veritable prospects for future survival. Although some synthetic polymers are biodegradable, the process of production that is usually non-green adds to environmental pollution. Natural polymers are naturally occurring organic molecules such as cellulose, starch, glycoproteins and proteins. They are mostly obtained from plant sources, but are also produced in animal and microorganisms. One of the most abundant natural polymers of multi-dimensional and multifaceted application is starch. Starch is used across wide-range applications spanning engineering, food and beverages, textile, chemical, pharmaceuticals and health, etc. This is because it can readily be modified into products of desired physicochemical characteristics, thus making starch a potential tool for industrial and economic revolution. The global trade balance for starch and derived products is about \$1.12 trillion, presenting a huge opportunity for more investment in starch production. Africa's negative starch trade balance of about \$1.27 trillion makes it a potential investment destination for starch production. This chapter discusses the use of starch in various industrial sectors, its potentials for sustainable economic development and as a veritable natural polymer for economic revolution.

Keywords: starch, natural polymers, environmental protection, green economy, industrial uses, sustainable development, economic revolution

1. Introduction

The major form of stored energy as carbohydrate in plants is Starch. It is a naturally occurring biopolymer consisting of a mixture of highly branched amylopectin and linear amylose residues. The two alpha-glucan residues make up 98–99% of the total net weight of starch [1]. Amylopectin residue is made up of linear D-glucopyranose chains linked by O-α-(1 → 4) glycosidic bonds and branching occurring as O-α-(1 → 6) glycosidic bonds. Amylopectin biopolymers are brittle. On the other hand, amylose residue has O-(1 → 4)-α-Dglucan linkages and is film-forming [2].

Naturally occurring starch has limited industrial applications its poor functionality such as poor water solubility at room temperature, retrogradation of its paste or

IntechOpen

gel, texture and taste. The functionality of starch can be modified through physical, chemical and/or genetic processes [3]. Due to the reactive nature of its monomers which is mainly as a result of their free hydroxyl (–OH), starch is easily modified to attain required functionalities for industrial purposes. For instance, if heated to high temperatures and in the presence of a plasticizer like glycerol, it exhibits comparable melt and flow characteristics as regular synthetic thermoplastic [4]. Undesirable characteristics such as hydrophilicity and low tensile strength are mitigated by introduction of hydrophobic fillers and materials that could enhance tensile strength. Likewise, different materials are used to improve thermal stability, plasticity and mechanical strength required in packing/packaging materials Advancement in material science which created thermoplastic starch has made starch a veritable resource with tensile applications in packaging and mechanical parts [5].

The ability to modify starch into biomaterials of different functionality has made it one of the most versatile and renewable natural polymer in existence. As a major type of food, it could be modified to enhance flavor, texture, thickness, taste, stability and/or shelf-life. Thus, it has found industrial application in food and beverage industry as food products or additives for enhancing the texture, stability, shelf-life and quality of products. It malleability with new technologies such as nanotechnology has expanded its scope of application in health and pharmaceutical and cosmetics. Starch could be used as both excipients and drug delivery vehicle [6].

The starch industry is at the very heart of food production: supplying hundreds of ingredients for use in thousands of food products and animal feed. At the same time, starches play a vital role in a wide variety of products beyond food. Natural and modified food starches can be found in products and processes in the consumer products, pharmaceutical, energy, industrial and chemical sectors. With the world beginning a gradual shift away from fossil fuels as the primary engine of economic prosperity, there will be a larger opportunity for starch producers to contribute renewable, sustainable materials through the bioeconomy [7]. This chapter discusses the various industrial application of starch which if exploited economically could provide strong foundation for economic revolution.

1.1 Starch sources

Starch is the second most abundant renewable bioenergy resources after cellulose, with an estimated global production of over 56 million tons per annum since 2006 [8, 9]. A variety of plant serves as the sources of starch consumed by humans. Starch storage in these plants occurs in grains or root tubers. Although the list of these plants which are either cultivated or found in the wild are endless, the major sources of food starch include corn, cassava, sweet potato, wheat, and potato. Sorghum, barley, rice, millet, yam etc. serve as minor sources in different parts of the world [10, 11]. A huge number of unexploited sources exist and majority is in the wild. **Table 1** shows the starch contents of some of these sources.

1.2 Contemporary application of starch and its derived products

By 2050, the population of the world is projected to exceed 9 billion and the demand for food is expected to rise by 70%. With growing environmental concerns and concept of green economy, reliance on fossil resources for energy and raw materials for industrial use etc. has attracted critical evaluation, and the prospect of a new trajectory for bio-based raw materials has become more imminent. Starch which

Source	Plant storage part	% starch content	References
Soft wheat	Cereal grain	77.90	[10]
Hard wheat	Cereal grain	77.40	[10]
Waxy rice	Cereal grain	74.76	[10]
Millet	Cereal grain	70.00	[12, 13]
Fonio millet	Cereal grain	68.00	[10]
Sorghum	Cereal grain	67.70	[10]
Cassava	Root	65.71	[10]
Taro/cocoyam	Corm	63.74	[10, 14]
White yam	Tuber	58.02	[10]
Rye	Cereal grain	58.00	[10]
Lesser yam	Tuber	54.70	[10]
Barley	Cereal grain	53.60	[10]
Sweet potato	Root	52.54	[10]
Yellow yam	Tuber	41.72	[10]
Potato yam	Tuber	38.10	[10]
Sweet corn	Cereal grain	36.23	[10]
Water yam	Tuber	31.90	[10]
Bitter yam	Tuber	20.48	[10]

Table 1.
Starch sources.

is one of the most abundant and affordable natural polymer poses a more reliable and sustainable substitute to the non-renewable, non-green, and exhaustible fossil sources. This opens prime opportunities for starch-based bio-economic revolution especially for countries with the right cultivation and production technologies for starch sources and starch-based products, respectively [7].

Starch can be used as biopolymers in many ways including as a raw material for human foods and animal feeds/feedstock, as bioethanol for food and fuel, as particulate filler and adhesive in paper and textile sizing, as well as bioplastics in packaging materials (**Figures 1** and **2**) [8, 9, 15]. It is also deployed in a wide array of other consumer goods in health and pharmaceuticals, and chemical sector. Corn starch product is used in 3-D printing inks, and emerging reports indicates its potential for nanomedicine technology as a tool for delivering treatments to specific sites. Some other categories of products include starch-based detergent products, starch-based binders, starch in biodegradable polymers, starch-based products for pharmaceuticals and cosmetics, and starch hydrolysates for fermentation [15].

1.3 Foods

The food industry is very large and diverse, and includes the raw unprocessed food and the processed and modified ones. Central to world food production and sustainability is the starch industry. Native starch is eaten in unprocessed raw or cooked form as grain or cereal meals, and flour dough or mashes with soups and other forms of condiments, in many developing countries. In West Africa, cooked

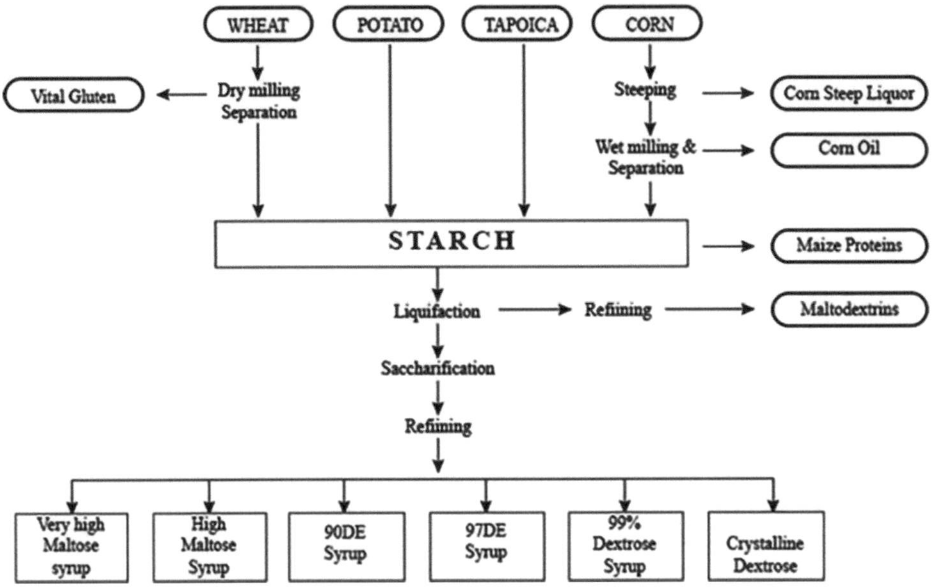

Figure 1.
Fermented and refined products from starch. (source: [15]).

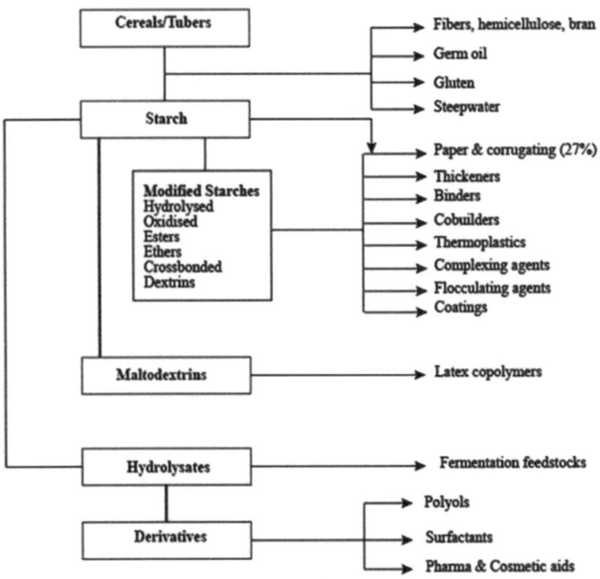

Figure 2.
Industrial application of starch. (source: [15]).

and mashed cassava (*fufu*) or fried cassave (*garri*) meals, corn grains and other starches from millet etc. are the bases for food security.

The food industry is very mindful of safety of chemical residues hence not all types of native or modified starches are used in the foods. Some modified starches are used as binder in assaulted foods, ready-made meat and snack seasonings. Others are used as anti-sticking agents and dustings for chewing gum and bakery products, crisping coating for fried snacks, fillers to replace fats and in sauces or creams to enhance lusciousness in ice cream and salad dressings. Modified starches are also used as flavor encapsulating agents and emulsion stabilizers in beverages. They are used as creamers, in canned foods, foam stabilizer in marshmallows, gelling agents in gum drops and jelly gum, and as expanders in baked snacks and cereal meals [2]. Starch derived products are used for the production of animal feeds [15]. About 10–15% of corn produced in the US is processed annually for starch derived products by corn refiners. These starch derived products are used in across the food, beverage, healthcare, pharmaceutical and other sectors. This has a dominant multiplier effect on the United States economy [7].

2. Sweeteners (syrups and sugars)

The CRA 2019 report put US production volume of sweeteners from corn starch refining for 2018 at 14.45 mMt d.w. These include glucose syrups, high fructose corn syrups maltodextrins, dextrose, corn syrup solids etc. [16]. The many other forms of starch are now been adopted as substitute raw material for corn. These products find applications in foods, beverages and pharmaceuticals for taste, flavor, color and texture enhancements.

2.1 Bioethanol

Ethanol is one of the most important organic solvent in the chemical industry. It is also the basic raw material for the wine, brewery and beverage industries. Due to energy fuel sustainability concern, the world is focusing on renewable energy include energy for renewable biological materials like starch. According to the World Integrated Trade Solution (WITS) data of the World Bank, the global volume of ethanol export in 2020 was over 14.2 billion liters, with a net weight of over 13.3 billion kg, valued at over $10.2 billion USD. While the volume of imports was 9.3 billion liters of 9.2 billion net weight and valued at $10 billion USD, with the top 10 importers as Netherlands, European Union, Germany, USA, Canada, Japan, Brazil UK, France and Korea Republic [17].

The production of ethanol from biological materials such as starch has gained interest in recent years because of the low-cost raw materials, starch, and the uncomplicated process involved. These processes, either chemically or biotechnologically based, are environmentally friendly. Bioethanol is already being used in many countries as octane enhancer for gasoline to produce gasohol. Starch from corn and cassava can be used for ethanol production. Due to concerns on quality, corn starch maintains a premium as the primary source of food and pharmaceutical grade starch and may not be very feasible for use as biofuel based on demand. However, sources like cassava may offer a ready source. Nigeria is currently the world's biggest producer of cassava starch although almost 100% of it is consumed locally as food. The rising demand for biofuel offers countries like Nigeria opportunity for economic revolution in the production of biofuel from cassava. Adeleye et al. [18] has report an ethanol yield of 1.5 L of 78% (v/v) from 2.5 kg wet weight of cassava. Countries like South Africa and China are already at advance stages of developing cassava plantations for production of starch for industrial uses [19, 20]. According to the 2019 Corn Refiners Association (CRA) report, the US produced about 6.06 billion liters of Ethanol, and about 3.37 mMt of Starch in 2018 [21].

2.2 Adhesives

Adhesives are mostly used in wood panels production, leather works and paper and packaging. The global volume of adhesive consumed annually is over 3 billion kilograms, mostly from petroleum derived feedstock and other synthetic materials. Due to environmental and safety concerns, a lot of studies have been ongoing to develop bio-material based hot melt adhesives (HMAs) from starch and its modification derivatives, (poly)lactic acid, soy protein, lignin and tannin [22].

Wood panels are composite products made by bonding wood particles or fibers with adhesive binders to form a board, which may be medium density fiberboard (MDF) or high density fiberboard (HDF). These panels find applications in home, office and industrial building construction. Wood panel makers currently almost depend exclusively on formaldehyde and amino-based adhesives with high formaldehyde emission, and polymeric 4,4-diphenylmethane diisocyanate, which are synthetic products. Concerns for indoor emission of formaldehyde, a known human carcinogen, led to the development of low-emission melamine-fortified urea-formaldehyde adhesives and other adhesive that exclude formaldehyde. Urea has been the primary formaldehyde scavenger for wood-based panels. Environmental consideration has raised interests for more green adhesives from biodegradable polymers like starch, lignin, tannin and protein. Organic scavengers like tannin powder, charcoal and wheat flour have shown promising potentials in reducing formaldehyde emission. Bio-based adhesive for industrial use are few and expensive. Tannin and starch adhesive, soy protein based adhesive and lignin-based adhesive are available for limited application in panel production [23].

Adhesives from native starch rely on hydrogen bonds which is weaker than chemical bonds. Due to their hydrogen bonding, they easily bond with water molecules and are therefore readily soluble and not-water resistant. Crosslinking starch produced with synthetic reagents such as epoxy chloropropane, sodium borate, hexamethoxymethylmelamine, formaldehyde, and isocyanates, tends to give better bonding force and water resist [23, 24]. Although no economically viable bio-based crosslinker reagents for starch are available, research on hot melt adhesives prepared by crosslinking modified propionyl starch with glycerol and polyvinyl alcohol (PVOH) has shown improved tensile strength of up to 2.0 MPa. Hence starch still offers a viable potential for future researches in 100% bio-based adhesive for wood-based panels industry [22].

The global volume of production and consumption of hot melt adhesive is on the increase and is about 15–21%, with an annual consumption growth rate which is 1.5–2 times higher than other adhesives [22]. This presents a huge market potential for starch-based HMAs.

2.3 Pharmaceuticals and cosmetics

They separate corn kernels into their component parts to make hundreds of products that touch consumer lives in countless ways every day. For years, those ingredients have been used to make food taste better, cosmetics last longer, pharmaceuticals easier to swallow and plastics environmentally-friendly [7].

The fundamental physiochemical and functional properties of natural starches for instance their good biodegradability and safety, make them suitable for a wide array of health and pharmaceutical applications. Several types of modified starch polymers and their application in bone tissue technology as bone tissue engineering scaffolding [25, 26], drug delivery system as biodegradable nanomedicine-carrier based delivery system and implants [27, 28], and hydrogels have been studied by different scientists over the past few years [6]. Starch has also been demonstrated as a viable material for capping of nanoparticles from different metals like Au, Ag and Pt, because of their bio-tolerance and cost effectiveness [29]. It has also been demonstrated to have potentials for use as nanoparticles to stabilize emulsions, Pickering emulsions, which are useful in cosmetics, pharmaceuticals and foods [30]. Also, pharmaceutical grade starch from corn is use as coating and filler excipients in tablets and caplets as well as syrups in many pharmaceutical products. It is also applicable as disintegrating agents, carriers, lubricants, matrices for controlled release [15]. For its good qualities of being odorless, decolourisable, environmentally biodegradable and skin friendly, it is used in cosmetics and beauty products as emollients, humectants, thickeners, film forming agents and emulsifiers [15]. Many sources of natural starch have been study and found to be effective for the production of pharmaceutical grade starch.

Starch and its modified derivatives have been used in medicine as biodegradable films, inexpensive cure for athlete's foot, anti-sticking agents, relief rashes caused by prickling heat, relief skin itches caused by shingles, relieves rash caused by baby diapers, wound dressing and bandages and used to treat gastric dumping syndromes in children. It is considered a safe alternative to cancer causing talcum baby powder, and used to remove excess oil from scalps and relieves itching in children [31].

3. Thermoplastics and bioplastics

Plastic pollution and the global push for a more sustainable environment towards eliminating the use of non-biodegradable materials and reduction of hazardous emission form toxic materials has refocused the worlds efforts towards the use of biodegradable plastics based on natural polymers such as starch, cellulose, lignin and chitosan [32]. The current global annual production of plastics is estimated at about 368 million tonnes, out of which about 1% is bioplastics. It is estimated that bioplastic production capacity will increase from 2.11 million tonnes in 2020 to about 2.87 million tonnes in 2025 [33].

The starch-based bioplastics thermoplastic starch (TPS or TS) is produced obtained from gelatinize starch and plasticizers which is subsequently tuned pellets by extrusion is useless as a material. It can also be produced from polymerization of polylactic acid obtained from starch-derived sugars fermentation. Bioplastics from starch can be used for producing compost bags and disposable plastic household wares [32]. The desire for more ecofriendly plastics gives a huge economic prospect to the growth of biodegradable natural polymer-based bioplastics market [33].

3.1 Carbon/carbonaceous foams

Carbonaceous foams (CFs) are used all over the world and find application in military, industrial and domestic use, such as heat exchanger, electrode materials, catalyst carrier, adsorption, vibration damping and impact or sound absorption, electromagnetic shielding, radar absorption, filtration, and aerospace material, etc. Carbon/carbonaceous materials are generally made from non-renewable raw materials like coal tar pitch, mesophase pitch and synthetic materials, at high temperature (above 1000°C) and high pressure (in MPa). Although other materials such as sucrose and tannin have been used, more cost effective biomaterials have been successfully investigated. A more energy efficient process of producing CFs with excellent compressibility and mechanical strength has been demonstrated by using starch as raw material. The production requires much lower temperature (<500°C) and lower pressure (about 190 Pa) than conventional approaches [34]. The world annual production of polysaccharides is in excess of 150,000 million tonnes. But the production of CFs from polysaccharides such as starch is still very much at low level. This gives good opportunity for economic exploitation.

3.2 Other industrial uses

In the industry, starch powder is used in textiles, paper, inks and paints. They serve as fast absorbent polymer in water treatment, as cleaning agent in detergents, as desiccants to prevent mildew from ruining paper documents in storage, as fabric stiffener and yarn sizing, remove wax from wooden furniture, and as organic pesticide [6, 15]. Starch based alkylpolyglucosides (APG) are used in detergents with superior skin compatibility.

3.3 Economy of starch

The economy of the starch industry largely depends on the availability of sufficient volumes of raw materials and the value of the so-called co-products. During corn starch processing, for example, all components of the maize grain are valorized: after steeping and coarse crushing the germ is separated and yields the valuable maize germ oil while the steeping water is concentrated and sold as nutrient for fermentations. The oil press cake together with the corn gluten (protein) and the hulls (fiber), which are separated after fine grinding in additional refining steps from the pure starch granules, are utilized as components in animal feed. The starch as the main product is either dried and sold as native starch, or chemically modified to make it more suitable for more demanding end uses, or hydrolyzed to yield refinery products such as hydrolyzates (dextrin), glucose syrups, and high fructose syrups [31].

In wheat starch processing, the value of the vital gluten is an essential source of income and could be regarded as the head product. In contrast to cereal and pulse starch production, the extraction of root and tuber starches does not deliver co-products of comparable value. As the processes for the extraction are different for the mentioned crops the starch industry cannot easily switch from one source to the other in order to adapt to fluctuating market conditions both on the raw material and on the end products.

Statistics on starch production and export are only available for the US, UK, the European Union and a few other countries. According to a 2014 report, Africa consumes the least starch per capita, and accounts for just about 2% of the global consumption market of starch and its derivatives. The two leading raw materials, cassava and maize are mainly consumed as food. Production of derived products is concentrated in few countries, Nigeria, Egypt and South Africa, with South Africa in the lead. While maize is the general source of starch for food in Egypt and South Africa, cassava starch plays an equally significant role as food in Nigeria and the West African region [19, 35, 36].

Importation and cultivation of *Manihot esculenta* for Cassava starch in commercial quantities for production of derived products for industrial use is now a major trade in

Asia countries of Thailand and China. A 2018 FAO report put the global production of cassava at 277 million tonnes (fresh root equivalent) and was project to rise by 0.5% annually. The 85.7 million tonnes production from Africa represents about 31%. Only 27.7 million tonnes were traded in same year. Wheat production and trade was 727.9 million tonnes and 173.2 million tonnes respectively. No country in Africa is among the top 10 importers and exporters of starch products [37].

The starch consumption per capita for 2012 was in the order of Africa < South America < Asia < Europe < North America. Country-wise is Nigeria < Egypt < Mexico < Turkey < Russia < UK < South Africa < China < US. Africa also consumes the least sweetener per capita. This is followed by Asia, South America, Europe and North America. Africa consumed less than 1 kg per person while North America consumed over 40 kg per person in 2012 [36]. This scenario has not changed significantly. South Africa produces over 280,800 tons of corn starch annually and is experimenting on local cultivation of *M. esculenta* for local production of cassava starch [19]. It is estimated that about 72% of Africa's population have yet to participate in the formal starch and syrup market. Apart from few cassava production plants, ethanol plants are springing up in Nigeria that targets cassava starch as raw material for production of ethanol for industrial use and biofuel.

According to the International Federation of Starch Association (IFSA), starch and its derived products accounts for an annual revenue of $47.50 billion in the United States in 2020, and supported about 167,786 jobs to the tune of about $10.01 billion wages. It accounted for about $1.91 billion worth of US export with the value of starch only products amounting to about $339.62 million [7]. The overall industry impact on US economic output (direct and indirect impact) was estimated by the Corn Refiners Association (CRA) to be about $7.16 trillion annually with an average growth rate of about 6% between 2017 and 2019 (**Figure 3**), while the value of impact on export was about $149.19 billion annually with an average growth rate of about 0.77% (**Figure 4**). However, the industry impact of export actually grew at 4.5% between 2017 and 2018. The impact on jobs (**Figure 5**) and wages (**Figure 6**) was actually higher in terms of growth. Capital investments in starch manufacturing were about $20.62 billion and $20.28 billion annually in 2018 and 2019, respectively [16, 21]. The US data presents a good indication of the potential of starch-economy as an economic revolutionary tool

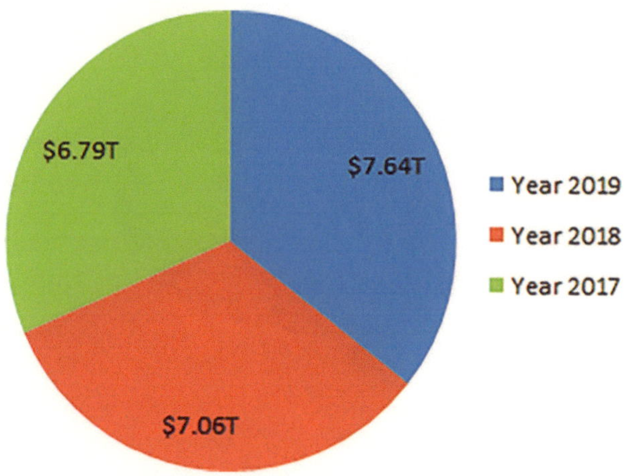

Figure 3.
Output economic impact of starch in the US, 2017–2019. (source: [16, 21, 38]).

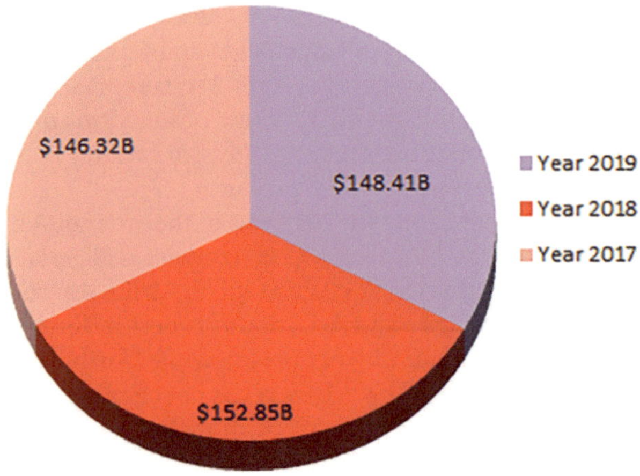

Figure 4.
Starch and derivative export value of the US, 2017–2019.

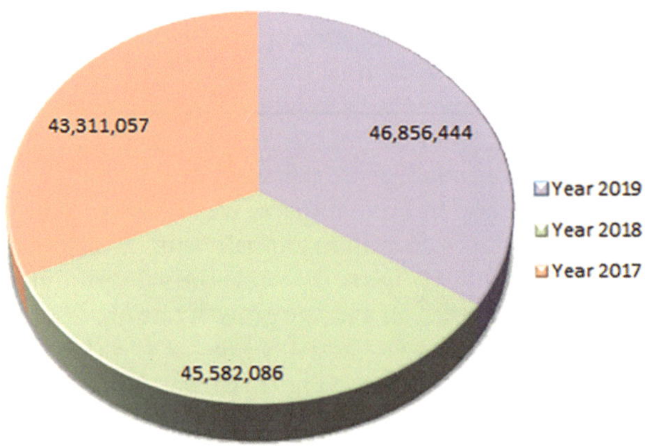

Figure 5.
Impact of starch industry on jobs in US, 2017–2019.

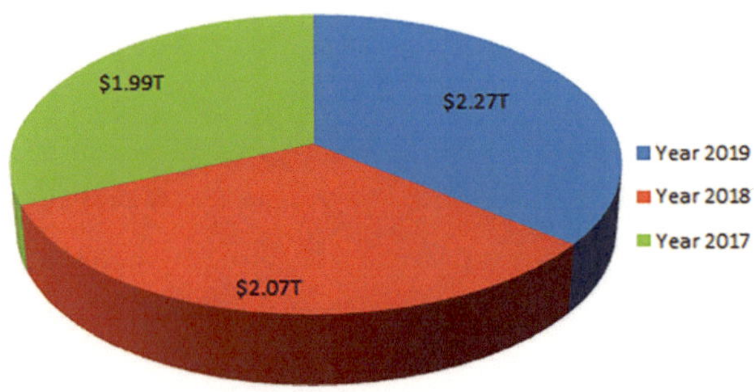

Figure 6.
Impact of starch industry on wages in the US, 2017–2019.

for job creation and export revenue. This potential is estimated to be higher for developing economies with higher opportunities for more investment in both upstream and downstream starch products value chain.

In 20 of the 28 EU member states, the European starch industry's 28 member companies process 24 million tons of EU agricultural raw materials into 11 million tons of starch-based ingredients and five million tons of proteins and fibers. Of starch-based ingredients, approximately 60% go to the food and beverage industry, and 40% to industrial applications (mainly to the paper, cardboard, pharmaceutical, and chemical industries as a renewable alternative to fossil fuel ingredients). Of the proteins and fibers, approximately 90% go to the animal feed industry and 10% to the food industry. The EU has 75 producers or plants and rely on Maize, wheat and starch potatoes, barley, rice, peas as feed stock to produce Native maize, wheat and potato starches, modified starches, maltodextrins, glucose syrups, dextrose, glucose fructose syrups, polyols, wheat gluten, other proteins. The industry supports 15,000 direct jobs, supports 100,000 indirect jobs, exports 1.6 billion Euros, and the annual industry turnover is 7.4 billion Euros [7]. The EU uses starch and starch products for confectionery, drinks, processed food, animal feeds, corrugating and paper making, pharmaceuticals and chemicals, as well as other non-food applications (**Figure 7**).

The Turkish starch industry contributes to local farming by processing 25% of locally grown maize from thousands of local farmers each year. The industry separates its starch from corn, drying it to native and modified starches or breaking it down to its sugars and other value-added products. Almost 100% of corn kernels are converted to economically valuable products. The industry supplies edible oils, fish, calf, lamb and poultry meat, paper, textiles and more to local and international food and beverage industries. With its wide-ranging portfolio, from basic to high-end products, the Turkish starch industry is competitive and has the capacity to grow. There are 9 producers which use maize as feed stock and produces Glucose and fructose syrups, native corn starch, modified corn starches, crystalline fructose, polyols, maltodextrins, corn gluten and feed, and ethanol. The industry supports 1,900 direct jobs and export over 400,000 tons annually.

Mexico has an evolving industry with annual output of between $300 million and $500 million USD. The industry supports 2,500 direct jobs and 7,500 indirect jobs. The brewing and paper markets depend largely on corn-based starch. The industry is growing due to investments in brewing capacity [7].

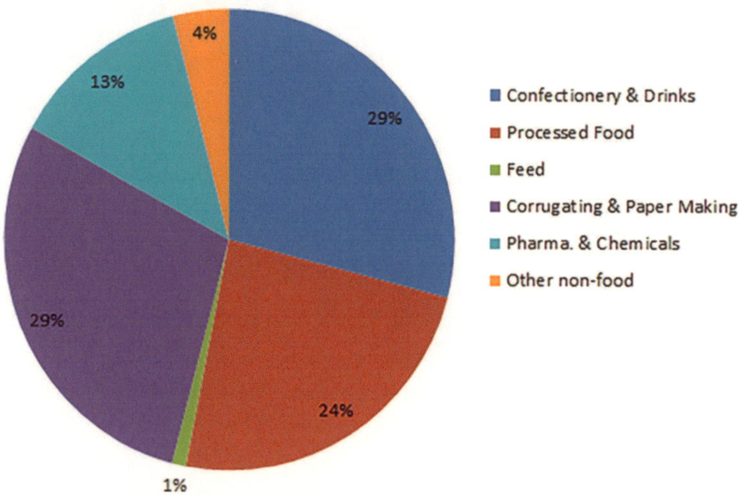

Figure 7.
Uses of starch in the European Union (source: [15]).

Founded in 1984, the China starch industry association has 280 members. Revolution in starch development has rebound in progressive development in the food, medicine, biology and chemical industries. It has also significantly contributed to growth of the national economy especially in agriculture where it has helped to sustain livelihood of local farmers through agro-economy. The annual industry output for starch and starch-based deep-processing products is estimated at 30.1 million tons and 16.3 million tons, respectively, from 170 producers. Major products include native starch, pregelatinized starch, chemically modified starch, starch sugar, polyols and ethanol, from feed stocks of corn, potato, cassava, sweet potato and wheat [7].

Starch development in Russia has put the country in an upward trajectory for economic revolution. For instance, the industry has 30 starch enterprises and 23 production plants. Ten of the Russia's 23 plants are responsible for 90% of all starch products from Russia. Annual production of starch and derivatives rose from less than 180,000 tons in 2013 to over 1.3 million tons in 2020. 70% of this production figures were sweeteners (**Figure 8**). The industry has a cumulative average growth rate (CAGR) of 8.65, 8.69 and 8.65 for starch, sugary products and total starch-based products, respectively. Corn accounts for about 800 thousand tons, wheat about 500 thousand tons, and potatoes about 30 thousand tons. The industry has invested about $358.4 million between 2013 and 2020 (**Figure 9**). **Table 2** provides Russian industry statistics on starch production and derived sugary products as obtained from Russian Federation Starch Union (RFSU) for 2013–2016 [39]. The 2015 import and export data is as provided in **Table 3**. The industry supports 4,000 jobs, has an annual output of $600 million and exports worth $28.5 million. The major products include native corn, wheat and potato starches, modified starches, glucose syrups, HFS, dextrins, maltodextrins, wheat and corn gluten, from the feed stock of corn, wheat, potatoes [7].

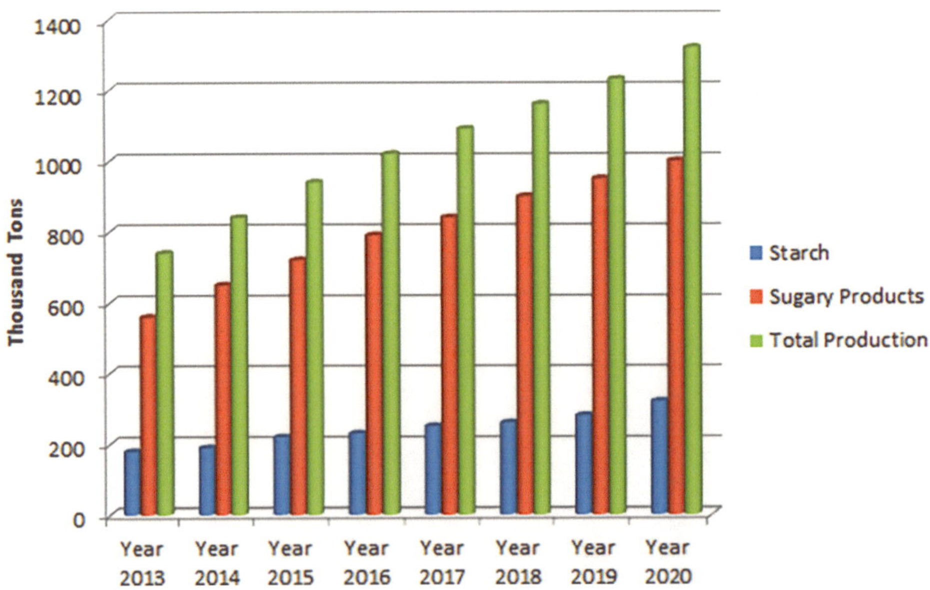

Figure 8.
Russian industry development indicators in the period 2013–2020 (according to the strategy for the development of the food and processing industry of the Russian Federation for the period up to 2020) [39].

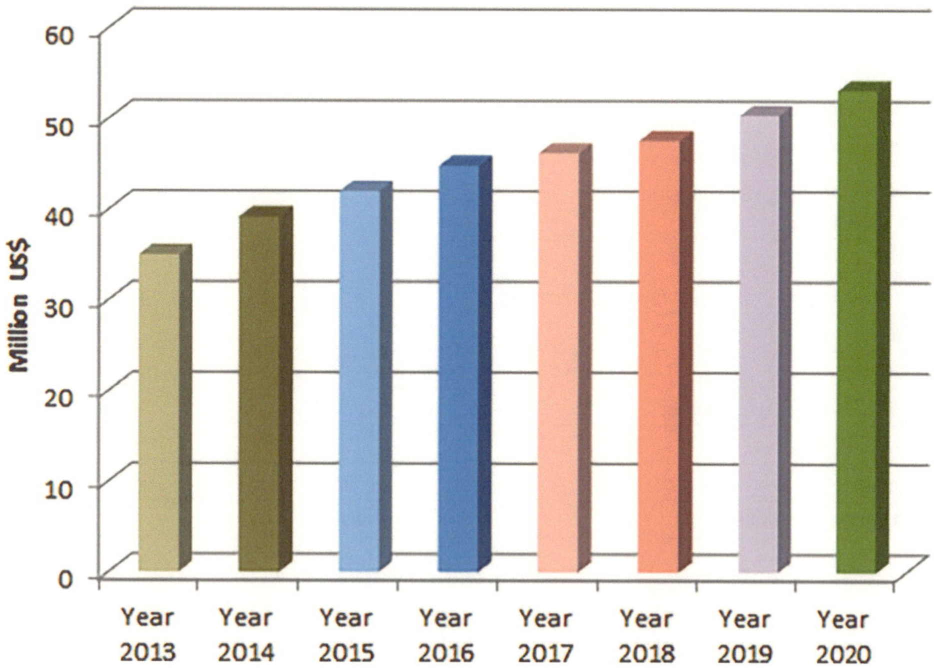

Figure 9.
The volume of investments in the development of the Russian industry for the period 2013–2020.

Tons	2013	2014	2015	2016
Dextrin	2547	2314	5216	7023
Starches, except modified	173,340	197,605	213,452	224,214
Starch syrup	372,383	173,340	437,033	464,349

Table 2.
Production in 2013–2016.

3.4 Global starch trade and opportunities

The value of starch and starch derived products imported and exported globally in year 2020 was $20,367,050,000 and $19,251,015,000, respectively according to figures from Trade Map [35]. African import and export were $1,921,266,000 and $652,628,000 (**Figure 10**), respectively, representing about 9.4% and 3.3%, of global figures respectively. This deficit represents a huge investment opportunity in starch production for the entire world. **Table 4** provides details of the starch products considered.

The high four years AGR of 10 for the global trade in flour (1105) and 7 for both cereal flour (1102) and inulin (1108) is a pointed to rising demands for these products. The imbalance in trade in wheat (1101) and malt (1107) for Africa expose the gap in export and import to Africa (**Table 4**). These rising demands and trade deficits could be bridged through more investments in starch production.

Products	Export in 2015 Rubles	Import in 2015
Corn Starch	12,538,346	11,404,606
Glucose Syrup (Patoca Starch Glucose)	8,812,834	7,699,115
Other Starch	0.988	141,900
Potato Starch	708,104	14,878,026
Maniok Starch	—	2,368,000
Rice Starch		159,900
Starches transformed into complex or simple ether	353,555	62,628,181
Wheat starch	341,147	3,181,534
Isoglucose	—	2400
Other, including syrup maltose	154,178	2,177,381
Dextrins	79,205	448,292
Other	50,720	1,919,326
Other starches modified	2,090,376	8,583,020
Other rice starch	300	—

Table 3.
Starch products export and import in 2015 in Russia federation.

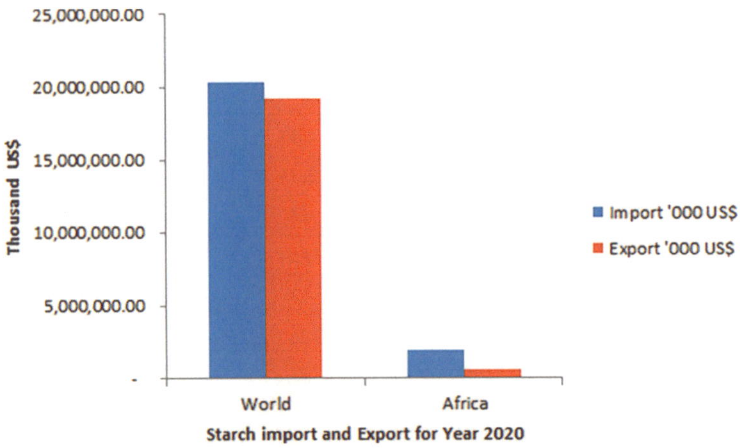

Figure 10.
Comparison of value of imported and exported in year 2020.

Products code	Imported by Africa	Exported by Africa	Trade balance—Africa	Import by The World	Exported by The World	Trade balance—World	*AGR 2016–2020
1101	615,881	302,215	−313,666	4,577,813	4,663,167	85,354	1
1102	164,983	61,057	−103,926	1,224,723	1,061,272	−163,451	7
1103	244,138	188,835	−55,303	1,228,918	1,141,939	−86,979	4
1104	95,193	15,921	−79,272	1,953,073	1,852,332	−100,741	6
1105	123,077	1598	−121,479	913,775	764,468	−149,307	10

DOI: http://dx.doi.org/10.5772/intechopen.102941

Products code	Imported by Africa	Exported by Africa	Trade balance—Africa	Import by The World	Exported by The World	Trade balance—World	*AGR 2016–2020
1106	11,303	15,724	4421	622,918	691,942	69,024	3
1107	471,918	24,550	−447,368	3,692,577	3,424,305	−268,272	1
1108	161,796	42,328	−119,468	4,712,232	4,348,129	−364,103	7
1109	32,977	400	−32,577	1,441,021	1,303,461	−137,560	0
	1,921,266	652,628	−1,268,638	20,367,050	19,251,015	−1,116,035	

*Products code: 1101 (Wheat or meslin flour); 1102 (Cereal flours excluding wheat or meslin); 1103 (Cereal groats, meal and pellets); 1104 (Cereal grains otherwise worked, e.g. hulled, rolled, flaked, pearled, sliced or kibbled; germ); 1105 (Flour, meal, powder, flakes, granules and pellets of potatoes); 1106 (Flour, meal and powder of peas, beans, lentils and other dried leguminous vegetables of heading); 1107 (Malt, whether or not roasted); 1108 (starches; inulin); 1109 (Wheat gluten, whether or not dried). *AGR means global annual growth rate for the products' trade between 2016 and 2020 (% p.a).*

Table 4.
Table of Africa and global starch trade in year 2020 in thousand US $ [35].

4. Conclusion

Starch is an abundant natural polymer with great industrial versatility. Various sources including maize, wheat, cassava, potato, rice and millet, abound in different countries of the world. Many countries especially those in Africa have not fully realized their starch production capacity and export potentials for international trade. The negative world trade balance of about $1.12 trillion for starch means opportunity for more investment in starch production. Starch and inulin with the highest world trade imbalance, and Malt, cereal and wheat gluten with relatively high world trade imbalance, could benefit more in future investment prospect (**Table 4**). A closer analysis shows that Africa has a higher negative trade balance of about $1.27 trillion than the world. As bad as this may look for Africa, it presents a huge opportunity for investments in global starch production. But Africa will have to do more to reduce the widening trade imbalance in starch products, especially in wheat and malt.

Conflict of interest

The authors declare no conflict of interest.

Author details

Obi Peter Adigwe, Henry O. Egharevba* and Martins Ochubiojo Emeje
National Institute for Pharmaceutical Research and Development (NIPRD),
Abuja, Nigeria

*Address all correspondence to: omoregieegharevba@yahoo.com

IntechOpen

© 2022 The Author(s). Licensee IntechOpen. This chapter is distributed under the terms of the Creative Commons Attribution License (http://creativecommons.org/licenses/by/3.0), which permits unrestricted use, distribution, and reproduction in any medium, provided the original work is properly cited.

References

[1] Haroon M, Wang L, Yu H, Abbasi NM, Abdin ZU, Saleem M, et al. Chemical modification of starch and its application as an adsorbent material. RSC Advances. 2016;**6**:78264-78285. DOI: 10.1039/c6ra16795k

[2] Egharevba HO. Chemical properties of starch and its application in the food industry. In: Emeje M, editor. Chemical Properties of Starch. London, UK: IntechOpen; 2019. DOI: 10.5772/intechopen.87777

[3] Ashogbon AO, Akintayo ET. Recent trend in the physical and chemical modification of starch from different botanical sources: A review. Starch/Staerke. 2014;**66**:41-57

[4] Quirino RL, Garrison TF, Kessler MR. Matrices from vegetable oils, cashew nut shell liquid, and other relevant systems for biocomposite applications. Green Chemistry. 2014;**16**:1700-1715. DOI: 10.1039/c3gc41811a

[5] Bodirlau R, Teaca CA, Spiridon I. Influence of natural fillers on the properties of starch-based biocomposite films. Composites Part B Engineering. 2013;**44**(1):575-583

[6] Palanisamy CP, Cui B, Zhang H, Jayaraman S, Kodiveri MG. A comprehensive review on corn starch-based nanomaterials: Properties, simulations, and applications. Polymers (Basel). Sep 22, 2020;**12**(9):2161

[7] International Federation of Starch Associations (IFSA). Washington, DC, USA: Corn Refiners Association. Available at: https://internationalstarch.org/#about [Accessed: 26 July 2021]

[8] Glittenberg D. Polymers for a sustainable environment and green energy. In: Matyjaszewski K, Möller M, editors. Polymer Science: A Comprehensive Reference. Vol. 10. Amsterdam, Oxford, Waltham: Elsevier; 2012. pp. 165-193

[9] Endres HJ, Siebert-Raths A. Polymers for a sustainable environment and green energy. In: Matyjaszewski K, Möller M, editors. Polymer Science: A Comprehensive Reference. Vol. 10. Amsterdam, Oxford, Waltham: Elsevier; 2012. pp. 317-353

[10] Raji AO. Utilization of Starch in Food and Allied Industries in Africa: Challenges and Prospects. London: IntechOpen; 2020. p. 24. DOI: 10.5772/intechopen.95020

[11] Horstmann SW, Lynch KM, Arendt EK. Starch characteristics linked to gluten-free products. Food. 2017;**6**(4):29. DOI: https://doi.org/10.3390/foods6040029

[12] Punia S, Kumar M, Siroha AK, Kennedy JF, Dhull SB, Whiteside WS. Pearl millet grain as an emerging source of starch: A review on its structure, physicochemical properties, functionalization, and industrial applications. Carbohydrate Polymers. 2021;**260**:117776. DOI: https://doi.org/10.1016/j.carbpol.2021.117776

[13] Mahajan P, Bera MB, Panesar PS, Chauhan A. Millet starch: A review. International Journal of Biological Macromolecules. 2021;**180**:61-79

[14] Coronell-Tovar DC, Chávez-Jáuregui RN, Bosques-Vega A, López-Moreno ML. Characterization of cocoyam (Xanthosoma spp.) corm flour from the Nazareno cultivar. Food Science Technology. 2019;**39**(2):349-357

[15] Röper HH. Industrial products from starch. In: Presentation at the 56th Starch Convention, Detmold, April 20-22, 2005. Cargill TDC Food, Cerestar R&D Centre, Vilvoorde, Belgium

[16] Corn Refiners Association (CRA). CRA Industry Overview 2019. Available from: https://corn.org/wp-content/uploads/2020/02/CRA-Industry-Overview-2019-Final.pdf [Accessed: July 29, 2021]

[17] World Bank (WB). World Integrated Trade Solution (WITS) Data. Available from: http://wits.worldbank.org/WITS/WITS/Results/Queryview/QueryView.aspx?Page=DownloadandViewResults HYPERLINK "http://wits.worldbank.org/WITS/WITS/Results/Queryview/QueryView.aspx?Page=DownloadandViewResults&Download=true"& HYPERLINK "http://wits.worldbank.org/WITS/WITS/Results/Queryview/QueryView.aspx?Page=DownloadandViewResults&Download=true"Download=true [Accessed: July 30, 2021]

[18] Adeleye TM, Kareem SO, Bankole MO, Atanda O, Adeogun AI. Ethanol production from cassava starch by protoplast fusants of Wickerhamomyces anomalus and Galactomyces candidum. Egyptian Journal of Basic and Applied Sciences. 2020;7(1):67-81. DOI: https://doi.org/10.1080/2314808X.2020.1746884

[19] Phaleng L. Trade analysis of the starch industry in South Africa (HS 1108). Trade Probe. 2017;(68):3-5

[20] Industrial Development Corporation (IDC). A study on the market potential for increased industrial starch production in South Africa. Urban-Econ Development Economists. Lake View Office Park Area 7, 137 Muckleneuk Street, Brooklyn; 2017 Available from: https://www.idc.co.za/wp-content/uploads/2018/11/APCF-Research-Grant_Starch_Final-Report-2017-October-2017.pdf [Accessed: July 28, 2021]

[21] Corn Refiners Association (CRA). CRA industry overview 2020. Available from: https://corn.org/wp-content/uploads/2021/02/CRA-Industry-Overview-2020.pdf [Accessed: July 29, 2021]

[22] Zhang Z, Macquarrie DJ, Clark JH, Matharu AS. Chemical modification of starch and the application of expanded starch and its esters in hot melt adhesive. RSC Advances. 2014;**4**:41947-41955

[23] Hemmila V, Adamopoulos S, Karlsson O, Kumar A. Development of sustainable bio-adhesives for engineered wood panels—A review. RSC Advances. 2017;**7**:38604-38630

[24] Qiao Z, Gu J, Lv S, Cao J, Tan H, Zhang Y. Preparation and properties of isocyanate prepolymer/corn starch adhesive. Journal of Adhesion Science and Technology. 2015;**29**:1368-1381

[25] Salgado AJ, Figueiredo JE, Coutinho OP, Reis RL. Biological response to pre-mineralized starch based scaffolds for bone tissue engineering. Journal of Materials Science: Materials in Medicine. 2005;**16**:267-275

[26] Gomes ME, Godinho JS, Tchalamov D, Cunha AM, Reis RL. Alternative tissue engineering scaffolds based on starch: Processing methodologies, morphology, degradation and mechanical properties. Materials Science and Engineering: C. 2002;**20**:19-26

[27] Liu C, Ge S, Yang J, Xu Y, Zhao M, Xiong L, et al. Adsorption mechanism of polyphenols onto starch nanoparticles and enhanced antioxidant activity under

adverse conditions. Journal of Functional Foods. 2016;**26**:632-644

[28] Xiao SY, Liu XM, Tong CY, Zhao LC, Liu XJ, Zhou AM, et al. Dialdehyde starch nanoparticles as antitumor drug delivery system: An in vitro, in vivo, and immunohistological evaluation. Chinese Science Bulletin. 2012;**57**:3226-3232

[29] Ban DK, Pratihar SK, Paul S. Controlled modification of starch in the synthesis of gold nanoparticles with tunable optical properties and their application in heavy metal sensing. RSC Advances. 2015;**5**:81554-81564

[30] Cazotti JC, Smeltzer SE, Smeets NMB, Dubé MA, Cunningham MF. Starch nanoparticles modified with styrene oxide and their use as Pickering stabilizers. Polymer Chemistry. 2020;**11**:2653-2665

[31] Ulery BD, Nair LS, Laurencin CT. Biomedical applications of biodegradable polymers. Journal of Polymer Science Part B: Polymer Physics. 2011;**49**:832-864

[32] Di Bartolo A, Infurna G, Dintcheva NT. A review of bioplastics and their adoption in the circular economy. Polymers. 2021;**13**:1229. DOI: https://doi.org/10.3390/polym13081229

[33] European Bioplastics. Bioplastics Market. Available from: https://www.european-bioplastics.org/market/ [Accessed: July 30, 2021]

[34] Lei H, Wu Y, Yang S, Fu C, Huo J. A simple strategy for converting starch to novel compressible carbonaceous foam: Mechanism, enlightenment and potential application. RSC Advances. 2018;**8**: 32522-32532

[35] International Trade Centre (ITC). Available from: https://www.trademap.org/Index.aspx [Accessed: July 30, 2021]

[36] Giraldello S, Ribeiiro M, Elrington J, Todd M, Bentley S. Starch and Fermentation Analysis. Oxford, UK: LMC International Limited; 2014. http://projects.nri.org/cassava-ipci/images/documents/LMC-SFA-2014-Mar.pdf [Accessed: July 28, 2021]

[37] FAO. Food Outlook - Biannual Report on Global Food Markets—November 2018. Rome; 2018. p. 104

[38] Corn Refiners Association (CRA). CRA Industry Overview. 2018. Available from: https://corn.org/wp-content/uploads/2019/02/CRA-Industry-Overview-2018-Final.pdf [Accessed: July 29, 2021]

[39] Russian Federation Starch Union. Industry overview. Moscow, Masha Poryvaeva: Domnikov Business Center; Available from: https://starchunion.com/obzor-otrasli/ [Accessed: July 27, 2021]

Chapter 4

Health Benefits of Starch

Teodoro Suarez-Diéguez and Juan Antonio Nieto

Abstract

In recent years, scientific research has focused on evaluating the relationship between consumption and the effect of food components on the body, with the aim of improving the health condition of the population. In particular, starch is the main component in grains and provides most of the energy in the diet. It is classified according to its nutritional characteristics as rapidly digestible starch (RDS), slowly digestible starch (SDS), and non-digestible starch (RS). Several studies have reported that different starch fractions show a correlation between digestibility and assimilation with physiological effect and metabolic impact. Each type of starch fraction consumed shows a different postprandial response, such that SDS and RS generate a slower absorption rate and lower serum glucose concentration, leading to a gradual uptake of glucose into the tissue, as well as a probiotic effect. Current reports suggest that consumption of SDS- and RS-rich products can generate a postprandial response of prolonged glucose uptake without hyperglycemic peaks, and improve the efficiency of modulation of carbohydrate metabolism. In this regard, there is a growing interest in carbohydrates with functional effects generating an emerging area of study. The aim of this chapter is to describe the potential functional effect and metabolic impact of consumption of the SDS and SR fractions of starch.

Keywords: starch, digestion, gradual energy, functional properties, health benefits

1. Introduction

Several research has been focused on investigating the relation between the intake of the different compounds existing in food and their health benefits [1]. A special emphasis has been paid to study their potential benefit effects on chronic and metabolic diseases, such as diabetes or obesity, in order to improve the life quality of the individuals with these pathologies [2, 3]. In this context, carbohydrates (CH) are the main macronutrient in food and contribute 45–55% of the required energy [4]. Specifically, starch is the main CH of the diet, and therefore, it provides the majority of the required energy. Starch is content on cereals, legumes, roots, nuts, and their derived products. During gastrointestinal digestion, starch is first hydrolyzed in the mouth by the activity of the salivary α amylase, able to hydrolyze the glucose-glucose bonds with direction α(1-4), releasing diverse dextrins and maltose [5]. The starch digestion is completed in the intestine by the digestive action of the intestinal enzymes α amylase, isomaltase, and glucoamylase, that provoke the starch debranching on the α(1-6) bonds and the hydrolysis of the α(1-4) bonds, releasing high amounts of glucose [6, 7]. These glucose are absorbed in the intestine by the

IntechOpen

Sodium-Dependent Glucose Transporter 1 (SGLUT-1), causing a glycaemia increase that provokes the release of insulin [8, 9]. However, the various botanical and industrial starches show different behaviors during the gastrointestinal digestion process, as a consequence of their different structural characteristics and physicochemical properties [7]. Therefore, regarding their digestion behavior, the diverse starch fractions can be classified as rapidly digestible starch (RDS), slowly digestible starch (SDS) [7, 10] and a crystallized starch fraction non-digestible denominated resistant starch (RS) [11–13]. The botanical origin shows a great influence in starch digestibility since it set up their structural characteristics and physicochemical properties, and therefore, the amount of each starch fraction [5, 7, 10].

The consumption of food rich in SDS is associated with a progressive release of glucose, maintaining a sustained energy source along the time compared with products with low SDS amounts and higher RDS contents [6, 10, 12, 13]. As consequence, the metabolic response generated by foods with higher SDS content shows a clear association with better postprandial metabolic parameters in healthy people but also in diabetic and obese individuals [13–16]. Thus, it is necessary to study and identify the mechanisms that relate the differences between the total glucose intake under starch form and the total absorber glucose after starch intake in order to understand the glycaemia response and metabolic profile of the different starches [14, 15]. The kinetic of the intestinal absorption of the released glucose can be used as valuable information to predict the postprandial changes in the blood glucose concentration and plasmatic insulin circulation [17, 18]. In this context, the diverse postprandial glycemic responses have been associated with the different starch fractions, where SDS and RS fractions show a slower glucose absorption rate, and therefore, attenuated glycemic response as well as less intense insulinemic responses [15, 19, 20]. Since high glycemic and insulinemic responses are associated with chronic diseases, mainly with the development of type 2 diabetes, a growing interest in the study of the CH and their metabolic responses exists, principally focused on the CH associated with a lower and maintained glucose absorption and therefore a mitigated glycemic response, characterized by the lack of hyperglycemic peaks and a maintained provided energy [19, 20].

2. Digestibility

Starch is largely digested among the gastrointestinal tract, being hydrolyses and absorbed at least 75% of the intake molecules [7, 21]. However, starch digestibility is conditioned by diverse factors, such as the acidity of the medium (a factor that reduces the activity of the amylase) or the cooking process, which gelatinize and solubilize the starch, increasing the accessibility of the digestive enzymes [4, 7, 9]. The starch digestion in the gastrointestinal tract occurs in diverse steps with the contribution of different digestive enzymes [8, 9, 16].

The first digestion step occurs in mouth digestion. Together with the reduction of the particle size of the intake of food by the chewing process, the release of salivary α amylase rules the digestion in the mouth. This enzyme shows a specific endo-hydrolytic activity on glucose-glucose α-(1,4) bonds but without activity on α-(1,6) bonds [10, 16, 22]. The endo-hydrolytic activity allows to hydrolyze the α-(1,4) bonds within the polymeric molecule, releasing lower glucose chains with direction α-(1,4), such as oligomers, maltotriose, or maltose [9, 18]. The mouth phase of the starch digestion is a short event because of the few time that takes the chewing and swallowing process, even though the first digestion products appears at 10–20 s after ingestion, increasing the digestion products as the mouth digestion progress. It is important to point that in contrast to other substrates, the starch size is higher than the digestive enzymes α-amylases, allowing many possible points within the molecule for the enzyme union, thus facilitating the enzyme activity [4, 9, 22].

The stomach digestion does not release specific enzymes with digestive action for the starch molecule. However, the swallowed saliva together with the alimentary bolus may exert a residual activity until reaching an acid pH able to inactivate the salivary α amylases. Consequently, the digestion of the complex CH can undergo total hydrolysis of 10–40% before reaching the small intestine [4, 22, 23]. During the intestinal digestion, starch is hydrolyzed by the activity of the intestinal α amylase, other endo-hydrolytic enzymes with activity on the glucose-glucose α-(1,4) bonds. However, this enzyme is not able to hydrolyze the α-(1,6) bonds existing in the branching amylopectin [10, 16, 18, 23]. Because of that, the activity of this enzyme is complemented with the activity of the isomaltase (able to hydrolyze the α-(1,6) bounds) and the glucoamylases, that hydrolyze α-(1,4) bounds mainly of glucose oligomers, such as dextrins. The combined action of these three enzymes becomes the successive digestion products, such as dextrins, oligosaccharides, and maltotriose, into maltose and glucose molecules. Besides, the action of the maltase enzyme of the intestine brush border becomes maltose into two glucose molecules [9, 16, 22]. The result of the complete digestion process of the ingested starch is a high amount of released glucose, being transported into the enterocytes by the transporter SGLUT-1 and excreted to the portal vein by the intestinal Glucose Transporter 2 (GLUT-2), provoking an increase of the glycaemia and an insulinemic response [14, 22, 23].

Starch digestibility is a factor of the food quality. The effect of the different starch fractions on the postprandial metabolism depends on the velocity and degree of the starch digestibility. SDS and SDS are categorized as glycemic starches and constitute the digestible starch fraction (DS). This DS fraction is hydrolyzed along the gastrointestinal tract whereas a fraction of the starch remains non-digestible, corresponding to the RS [4, 6–8]. The DS fraction is completely assimilated in the small intestine, responsible for the increase of the postprandial glycaemia [13, 14, 16, 24]. RS is characterized by a crystalline structure that avoids the digestibility of the molecule by the human digestive enzymes. As consequence, this fraction reaches the colonic tract, being fermented by the bacteria of the colonic microbiota, releasing short chain fatty acids (SCFA), such as butyric acid [11, 12, 25, 26]. The different digestibility found in the NS of the diverse botanical species has been explained as the interaction of various factors, such as the botanical source that condition the amount of each starch fraction, the starch granule size, the presence or absence of superficial pores on the starch granules, the relation amylose/amylopectin, the crystallinity degree on the molecule (correlated with the X-ray diffraction pattern) [24, 27], the association degree between diverse starch chains, the distribution and length of the branched chains of the amylopectin or the existence of interior channels and fractures on the starch molecule [24, 27, 28]. In addition, the industrial processing and cooking of starchy food can alter the starch properties, influencing the digestibility properties and also, the amount of each starch fraction [6, 7, 22, 29].

The rate and degree of starch amylolysis is determining factor in establishing the magnitude and duration of the postprandial glycaemic response. Currently, different in vitro tests have been considered to evaluate the rate and degree of starch hydrolysis as predictors of the physiological effect of food consumption [8, 17, 18, 23]. However, establishing the comparison of digestibility values found in the literature is a complicated task due to the variability in the methodology [8, 23, 27, 28, 30], as well as the variability of the type of enzyme used for the starch hydrolysis [16, 17, 28, 30]. In this context, the susceptibility level of the hydrolysis of retrograde samples depends on the type of α-amylase used (bacterial, fungal, pancreatic), enzyme concentration, hydrolysis time, and purity of the enzyme [17, 18, 28, 30, 31].

Among the aspects to consider studying and interpreting the in vitro digestibility of starch, and thus being able to establish postprandial physiological predictions, the understanding of the mechanics of action of digestive enzymes on starch hydrolysis should be studied. Currently, a kinetic model has been established that follows a pseudo-first order reaction for the analysis of starch hydrolysis, using as a tool the graph from the "logarithm-of-slope" (LOS) plot [17, 18]. This model allows conducting an analysis to be performed to classify the RDS, SDS, and RS fractions based on kinetic behavior in terms of a rate constant and the degree of hydrolysis. The in vitro hydrolysis level of starch is a frequently used indicator to determine the degree of total digestibility in starch samples. This parameter is represented by the equilibrium concentration ($C\infty$) at the end time of the kinetics of amylolysis, represented by the digestograms [16–18, 23]. Digestion rates are measured by the kinetic rate constant (k) (pseudo-1st order rate constants for starch amylolysis).

The kinetic constants of amylolysis in native starches (NS) and gelatinized, as in the case of cereals such as corn and legumes such as beans and broad beans, present a similar degree of level of enzymatic activity and affinity of the enzyme for the substrate [32]. In general, the gelatinized NS LOS plots of corn, beans, and broad beans show similar kinetic behavior. This implies that in the fastest phase the easily accessible starch is hydrolyzed, presenting a relative duration of 30 min, and later it simultaneously passes to a slower phase, which shows that the behavior of the graphics tends to be of a single phase. This implies that the easily accessible fraction is more available to the action of α-amylase, resulting in an increase in the degree of digestibility expressed as $C\infty$. The LOS graphs show a single linear phase hydrolysis process, considering a constant k with a similar behavior between the different varieties obtained by this method. Hydrolysis of gelatinized starches generally occurs in a single phase as gelatinization makes the starch fully accessible to the enzyme [33].

In the case of RS obtained by debranching and subsequent retrogradation of corn, beans, and broad beans, the LOS graphs show the behavior of an initial fast phase that inevitably has a prolongation. This implies that the reaction is characterized by a slower phase, and represents the fraction of starchless available, which is reflected in the values of k and $C\infty$ [32]. Thus, the slopes of the LOS graphs of the RS are notably lower than the NS, consequently, lower values of k are obtained in the RS samples compared to the NS samples that reflect a slow phase of hydrolysis, showing a lower affinity of the enzyme for the substrate.

Studies in NS of these varieties have shown that hydrolysis is faster in the first phase because the enzyme more easily accesses the starch fractions of the amorphous regions [16, 18, 34]. Type A starch, which is characteristic of cereal starches, shows a high proportion of short chains in the cluster, due to a large number of branches, which are more widely dispersed within the cluster, increasing the number of access points for amylase and substrate [24, 35, 36]. In addition, it has been considered that the crystallinity pattern of starches conditions their digestibility. NS with type A crystallinity patterns have pores and channels, whereby the enzyme penetrates into them and the hydrolysis reaction starts from the hilum region towards the outside of the granules, thus favoring the degree and speed of starch hydrolysis [35, 36]. On the other hand, NS with a type B crystallinity pattern does not present pores, showing a non-porous surface, thus giving a different hydrolysis pattern than type A, because the enzymatic digestion starts from the starch surface. This promotes the degree and rate of hydrolysis to be lower [24, 33, 37]. Likewise, it has been observed that legume NS with a crystallinity pattern of type C (A + B) shows a lower degree of hydrolysis. It has been reported that the difference in the degree of amylolysis in starches with type C patterns, is influenced by the presence of fissures or cracks on the surface of the granule that some varieties of legumes present. These cracks would allow the enzymes to have a higher diffusivity and to penetrate more quickly into the granule, hydrolyzing more easily the starch chains close to the granule surface [34, 36, 37].

This phenomenon would explain the difference in RDS levels between the different legume varieties. Likewise, the levels of SDS and RS depend on the degree of structural organization of the double helices within the crystalline lamellae and the distribution of these lamellae in the granule [38, 39]. A starch with higher SDS and RS content is characterized by a decreased level of susceptibility to hydrolytic enzymes and consequently lower digestibility, generating a moderate postprandial response without hyperglycemic peaks [13, 19, 20, 40]. This type of starch is the most suitable for consumption in the diet, especially in diabetic situations [13, 14, 19, 20]. In the case of RS, they have a lower k constant compared to the digestion kinetics of NS and represent a decrease in the catalytic efficiency of α-amylase due to the high degree of crystallinity of RS [11, 12, 25]. A low value of k is estimated to reflect a lower diffusion and a slower phase in the hydrolysis process by α-amylase, due to the degree of crystallinity of the different fractions of RS which inhibits the action of α-amylase [16–18, 28].

3. Nutritional characteristics of the starch

Starch is constituted by two different CH polymers, amylose, and amylopectin. Amylose consists of linear chains of glucose molecules linked by α(1-4) bounds, whereas amylopectin is constituted by linear chains of glucose molecules linked by α(1-4) bounds, with branching points linked by α(1-4) bounds every 15–30 glucose molecules [24, 35, 36]. Regardless of the botanical origin of starch, it is always composed of these two polymers, changing their amylose/amylopectin ratio in relation to their botanical origin [24, 35, 36]. Therefore, the differences in the nutritional characteristics of the diverse starches are a consequence of their bioaccessibility and bioavailability that depends on their composition on RDS, SDS, and RS.

3.1 Factors that influence bioavailability

The starch digestibility is a consequence of intrinsic factors, such as the starch morphology and physicochemical properties, and of extrinsic factors, determined by the physiological conditions of the individuals [6, 7, 9, 28], with a variability regarding genetic factors [41].

The intrinsic factors are referred to the structural and physicochemical characteristics. These properties are related to the granule and highly influenced by the botanical origin [6, 7, 9], or are the result of applied processes during the industrial processing or cooking processes. These factors condition the ability and accessibility of the digestive enzymes to starch molecule, influencing the hydrolysis degree [41, 42], and therefore, the intrinsic properties condition the bioaccessibility and bioavailability degree of the molecule [40–42]. Since these factors directly modulate the digestibility, condition also postprandial response [39–41]. Diverse processes modify starch assimilation. The grinding of the grain provokes ruptures of the starch molecule, enhancing the efficiency of the gelatinization and allowing a greater interaction of the digestive enzymes, increasing the starch bioavailability. The food products elaborated with grinding grains are characterized by faster starch hydrolysis and an immediately postprandial response [6, 7, 9, 34, 40].

The extrinsic factors are referred to the physiological conditions of the individuals that determined the bioaccessibility and bioavailability of the starch. These factors include age, gender, metabolic conditions, pathological alterations of the digestive tract, the efficiency of thc digestive process, gastrointestinal transit time, and genetic variations, among others [9, 22, 42]. Due to the genetic variation in the expression of the digestive enzymes [22, 41, 42] or because of specific physiological conditions, some individuals do not completely absorb the glucose release from the starch hydrolysis [6, 41, 42]. The food emptying from the stomach is a consequence of the gastric emptying, regulated by diverse factors such as the amount and volume of the ingested food, the type of macronutrient, the energy density of the intake, the particle size of the food matrix, the viscosity, the osmolality and the pH [7, 9]. Once the gastric content is released to the intestine, factors such as the viscosity influence the accessibility of the digestive enzymes to the chime and the nutrient releasing [9, 22]. In this context, the gastrointestinal transit time of the starch is inversely correlated with the amount of ingested starch and the contact time of the enzymes with the substrate [41, 42].

3.2 Classification of the diverse starch fractions

The digestibility and bioavailability of the starch vary depending on the intrinsic and extrinsic factors, closely related to the starch type and the botanical origin [6, 7, 9]. As consequence, starch can be classified according to their digestibility degree and releasing velocity of their constituent CH [6, 8, 10, 19, 21, 22, 43]. In this context, the diverse starch fractions can be classified according to these two parameters, the digestibility and assimilation degree, in three different fractions: rapid digestible starch (RDS), the slow digestive starch (SDS), and the resistant starch (RS) [6, 21, 28, 43]. All the starches are naturally constituted by an RSD and an SDS fraction, and a minor fraction of RS. The whole digestibility of the starch is a consequence of these three fractions and can be determined in vitro by measuring the glucose release of each fraction, as a predictor of their potential postprandial response [39, 40, 43].

The RDS is characterized to be the first completely digested fraction, showing a characteristic high velocity of glucose release, occurring the whole hydrolysis of this fraction within the first 20–30 min [6, 27, 28, 30]. The digestion of this fraction release oligosaccharide that are quickly hydrolyzed to glucose molecules [44]. This fraction is completely assimilated, being digested, and absorbed in the proximal duodenum [7, 24, 27, 28]. The behavior of this starch fraction during gastrointestinal digestion is a consequence of the amorphous structure of this starch fraction and the high gelatinization degree, being easily hydrolyzed by the digestive enzymes [6, 27, 28]. The RDS generate a fast increase of the glucose concentration in the blood after the starch intake and frequently provoke a hyperglycemic response. These highly fluctuant glucose levels generate increased stress on the regulatory systems of the glucose homeostasis with possible alterations or damages in cells, tissues, and organs [44]. Food with this characteristic (rich or enriched in RDS fraction) are the products with a high refinery degree, such as cereals flours used for bakery, pastries, biscuits, breath or fried foods, among others [24, 27, 29, 39].

Conversely, the SDS fraction is characterized by slow and progressive digestion, showing an intestinal absorption of the almost entire starch fraction. This starch fraction is composed of an amorphous but rigid structure, generating a more inaccessible structure that hinders the enzyme accessibility and an imperfect crystalline structure that limits the action of the digestive enzymes. As a consequence of this more complex accessibility for the digestive enzymes, the hydrolysis velocity is dramatically reduced, showing a slower digestion velocity [16, 39, 45]. The progressive release of glucose molecules resulting from its hydrolysis, allowing a progressive intestinal absorption of the glucose products, avoids high glycemic peaks, showing a lower but maintained postprandial response. This glycemic response is beneficial for healthy people but especially for diabetics [13, 14, 19, 30, 44]. The slow digestion and assimilation are related to several intrinsic factors of the starch molecule, as well as to structural changes that occurred during industrial processing or cooking [4, 6, 19, 27, 29, 44].

The structural properties of the food matrix may play an important role during starch digestion. When starch is contained internally in the food matrix, it plays an important role in reducing the velocity of the starch digestion because it get caught on the food matrix or by a barrier action of the rigid cell walls. This events are commons in legumes [7, 10, 27, 29, 45]. This coughed starch fraction is especially interesting for people with metabolic or chronic diseases, particularly type 2 diabetic or hyperlipidaemic ones [45].

RS is the starch fraction characterized by not be hydrolyzed by the digestible enzymes, reaching entirely the colonic tract [6, 11, 12, 21, 24, 25]. The European Research Project on Resistant Starch (EURESTA) defines RS as "the sum of starch and products of starch degradation not absorbed in the small intestine of healthy individuals" [11, 12]. When RS reaches the colonic tract, it is fermented into diverse SCFA, especially propionic and butyric acids, or conversely, it is eliminated through the fecal material [11–13, 25, 26]. Regarding their chemical nature, RS can be classified into diverse categories. **Table 1** summarizes the diverse types of RS currently identified and their main characteristics. Some RS occurs naturally in food, existing in the starch

Type	Source	Digestibility
AR-I. Physically inaccessible starch, encapsulated into the morphological structures of the grains	Whole grains and/or partially grinded raw grains	Poor grade and velocity of assimilation
AR-II. Granules of NS highly crystallized and non-gelatinized	Raw potatoes, bananas and legumes	Low digestibility and poor assimilation
AR-III. Crystallized starch o retrograded	Bread and baked goods	Low digestibility and poor assimilation
AR-IV. Modified starch with chemical processes (such as interlinking process or esterification)	Specific ingredients chemically developed	Resistant or partially digestible
AR-V. Amylose-lipid complex (type V) as consequence of food processing	Bread and baked goods	Resistant to the digestion process

Table 1.
Classification of the diverse types of RS.

granule, whereas others RS are the result of a structural reorganization on the molecule or the molecule interaction with other compounds, as a consequence of industrial processing, cooking, or intentioned chemical modifications [11, 12, 24, 25]. In any case, the RS formation is due to a structural reorganization of the amylose linear chains and their association with the amylopectin [25, 46, 47]. Accordingly, RS frequently occurs in starch molecules with a higher amylose/amylopectin ratio, increasing the available lineal chains of α-glucans to form crystalline and organized structures [24, 25, 46–48]. The molecular weight of these RS crystalline structures is close to 100 kDa.

4. Functional properties of the starch fraction

4.1 Functional properties of the SDS and RS

The functional properties of a compound are defined by the way to modify the metabolic patterns, generating a beneficial property. Diverse compounds present in plants show functional properties [2, 49]. In this context, the possible functional properties of SDS and RS have been specially studied in the last years because of their capacity to provide a maintained source of energy avoiding hyperglycemic responses, especially beneficial for diabetic patients. It is important to note that the attenuated glycemic response during starch consumption is dependent on the digestibility behavior that depends on the starch composition and physicochemical properties [50]. The influence on the variability in the glycemic response is considered as an important factor to control the glycemic response in diabetic and obese patients, frequently measured as the concentration of glycated hemoglobin (HbA1c) [51, 52].

Diverse investigations support the evidence of this beneficial effect of the SDS or RS fractions. Hasek et al. [53] studied the effect of the consumption of SDS using obese rats how experimental animal models fed with high fat diet (HFD). A group supplemented with SDS was compared with a group with RDS supplementation. The experimental group fed with SDS reduced the total daily intake of food compared with the RSD group. In addition, when evaluating the expression levels of (mRNA) of hypothalamic orexigenic neuropeptide Y (NPY) and Agouti-related peptide (AgRP), they observed that their expression was significantly reduced in the group fed SDS, as well as an increase in the hormone that produces anorexigenic corticotropin-releasing hormone (CRH). These researchers suggest that SDS can contribute to modulating the frequency of food consumption by activating the gut-brain axis, in addition to generating a reduction in the expression of genes of appetite-stimulating orexigenic neuropeptides and an increase in hypothalamic appetite-suppressing neuropeptide. Therefore, SDS may exert beneficial functional properties.

Breyton et al. [50] compared the effect of the consumption of a diet high or low in SDS in a group of diabetic patients. They observed a reduction in the variability of the glycaemia as well as the lower postprandial area under the curve (AUC) of the glycaemia when the high SDS consumption was considered. These authors consider that the modulation of the starch digestibility may be used as a useful tool for controlling the postprandial glycemic response in diabetic patients. Similar results were reported by Lambert-Porcheron et al. [54], that evaluated the postprandial response and digestibility of a product based on cereals with high or low SDS content in patients with overweight o metabolic risk. They observed a 2 h postprandial glycemic and insulinémica response lower in the group of patients that consumed the product with high SDS compared to the other group.

Considering the effect of RS consumption, diverse studies have reported beneficial effects on human health [11–13, 26, 50] (**Table 2**). In this context, the physiological effect of the RS consumption shows a great dependence on the biological origin, the total amount intake and the type of RS consumed [11, 12, 24–26]. In general, it can be considered that the RS consumption reduces the postprandial glycemic response and improve the glucose metabolism and homeostasis [11, 12, 14, 26] as a consequence of not being digested in the upper human tract [12, 45, 47, 50] and by contributing to a lower but maintained postprandial glucose in the diet [11–13]. These properties are especially beneficial for type II diabetic patients allowing better control of the glycaemia [11–13, 26], as well as for healthy individuals [12, 13, 26]. In addition, RS can exert a modulation effect on the satiety by modifying the secretion of adipokines and peptides responsible for this physiological process [12, 13, 26], such as an increase of the release of ghrelin, leptin, adiponectin, glucagon-like peptide 1 (GLP-1), peptide tyrosine (PYY) and gastric inhibitor peptide (GIP). The satiety promotion reduces the total food and calories intake preventing the excess of calories and the accumulation of fatty acids in the adipose tissue [12, 47, 50, 53]. Therefore, RS can be used as a useful tool to prevent and manage obesity [12, 13, 21, 26].

Other beneficial effect attributed to RS is its capacity to act as a prebiotic compound. Since RS is low or not digested in the upper gastrointestinal tract, it reaches the colon as part of the dietetic fiber [24, 25, 28, 46]. RS is fermented by the colonic microbiota bacteria releasing SCFA, especially butyric acid, an essential compound to maintain the intestinal epithelium permeability and associated with colon cancer

Physiological effect	Mechanism	Beneficial effect
Improve the insulinemic response	The lower digestibility reduces the glucose absorption and glycaemia response	Diabetes Insulinemic response Metabolic syndrome
Improve the intestinal epithelial health	Contribute to reverse neoplastic changes on the colon epithelium by the production of organic acids (butyric, propionic and acetic) as a product of RA fermentation.	Colorectal cancer Ulcerative colitis Intestine inflammation Intestinal constipation
Improves the lipid profile	Contribute to reduce the cholesterol and triglyceride blood levels	Cardiovascular disease Metabolic syndrome Lipid metabolism
Promote the satiety process	Reduce the total calories intake since RS is not digested in the upper gastrointestinal tract Promote the secretion of the hormones related with the satiety process	Obesity Diabetes
Increase the nutrient bioavailability	Improves the intestinal absorption of minerals like intestinal absorption of minerals like iron or calcium	Osteoporosis
Prebiotic effects	Promotes the growth of microbiota bacteria associated with healthy microbiota (*Bifidobacteiurm* and *Lactobacillus*)	Colon health Metabolic syndrome
Dietetic fiber effects	Reduces the atrophy of the colon epithelium, contribute to prevent the intestinal constipation	Improves the intestinal health Improves the intestinal peristalsis

Table 2.
Physiological effect of the RS consumption.

prevention. In addition, the SCFA can modulate satiety and lipid metabolism and are also substrates of the intestine gluconeogenesis. This process is associated with a reduction of hepatic gluconeogenesis and therefore a direct impact on the postprandial glycemic response [12, 13, 21, 25, 26, 50].

The prebiotic effect of the RS is associated with the RS fermentation by Bifidobacterias and Lactobacillus species allowing to promote the growth of this beneficial bacteria in terms of total amount and diversity [12, 25, 26]. Dysbiosis, defined as a negative alteration of the microbiota population, is generally associated with a reduction of the microbiota diversity and especially with the alteration of the Bacteroidetes and Firmicutes phyla, characterized in general by the increase of the Firmicutes/Bacteroidetes ratio and the reduction of the Bifidobacterias group. These microbiota alterations are associated with intestinal complications and the increase of the incidence of chronic diseases such as obesity, type 2 diabetes, cardiovascular diseases, and also with cancer [55–57]. The prebiotic effect of RS promotes the growth of Bifidobacterias and Lactobacillus species, contributing to maintaining a healthy microbiota, a continuous production of SCFA, maintaining the integrity of the intestinal epithelium, and as consequence, reducing the intestinal absorption of bacterial lipopolysaccharide (LPS) and therefore, reducing and preventing the endotoxemia, metabolic stress and chronic systemic inflammation, responsible or contributors of many chronic metabolic diseases [56, 57].

This prebiotic action of RS may exert an effect on glucose metabolism through different mechanisms, such as decreasing the gastric emptying, decreasing the glucose absorption in the intestine, favoring or promoting the production of GLP 1, as well as decreasing the expression in the transcriptional factors that intervene in the oxidation of fatty acids and lipogenesis, generating a decrease in the free fatty acids levels in the blood [56, 57].

4.2 Modulation of the glycaemia homeostasis by SDS and RS

As was exposed before, the nutritional characteristics of starch depend on its structural and physicochemical properties, responsible for the different digestibility, bioavailability, and postprandial glycemic and insulinemic response [4, 14, 15, 19]. Starches with higher SDS and RS fractions show a slower digestibility, progressive assimilation, and therefore a lower glycemic response. Conversely, high RDS contents are characterized by rapid digestion, assimilation and an elevated glycaemic response [19, 44, 50, 53, 54]. Therefore, the starch characterized by high SDS or RS contents can be used to avoid hyperglycemia [19, 26, 44, 45] and get a higher control of the postprandial glycemic response and glucose homeostasis in metabolic syndrome diseases, especially for diabetic patients [13, 54, 56]. The replacement of natural starchy foods by starches with higher SDS and RS content together with the commonly prescribed drug is an adequate strategy to get an integral control of the glycaemia response [4, 13, 15, 19, 26, 44, 47, 50, 54, 56].

Diverse in vivo studies show that RS consumption is associated with benefits in glucose homeostasis [13, 54, 56, 58]. Sun et al. [58] evaluated the consumption effect of RS-II in rats with type 2 diabetes, fed with a diet high in lipids and glucose for 4 weeks. The experimental group showed a reduction trend in glycaemia compared to the diabetic control group (without RS-II). Also, higher glycogen levels were determined in the liver and muscle of the experimental group compared to the diabetic control but similar levels that the non-diabetic group fed without RS-II. Similar results were determined by Zhou et al. [59] when RS from high amylose maize were evaluated during 4 weeks. The experimental group of diabetic rats fed with RS significantly reduced the glucose cholesterol and triglycerides blood levels compared with the control diabetic rats (non-fed with RS) and even HDL levels were determined twice as high in the first group. Zhou et al. [59] and Sun et al. [58] suggest that the biological mechanism responsible for the beneficial RS effects in the glycemic response is mediated by the promotion of the hepatic glycogen and the gluconeogenesis inhibition together with higher efficiency in the glucose intake by the muscular tissue [13, 59]. Gluconeogenesis is the metabolic pathway responsible for the endogenous synthesis of glucose molecules through the use of non-carbohydrates precursor molecules such as pyruvate, alanine, or glycerol. The phosphoenolpyruvate carboxykinase enzyme (PEPCK) converts the oxaloacetate to phosphoenolpyruvate during the gluconeogenesis pathway, whereas during the last step of this pathway the glucose-6-phosphatase (G6Pase) removes the phosphorous group in the 6 positions of the glucose-6-phosphate releasing glucose [13, 60, 61]. Diabetic patients are characterized by continuous activation of the PEPCK and G6Pase enzymes as a response to the low glucose intake by the muscle tissue as a consequence of an inadequate insulinemic response, causing hyperglycemia under fasted state [13, 61, 62]. In this context, diverse in vivo studies have observed a reduced expression of the PEPCK and G6Pase enzymes in diabetic rats treated with RS and therefore, a reduction of the gluconeogenesis activation [58, 59]. The glucose intake by the muscle tissue and the inhibition of gluconeogenesis provokes the activation of the AMP-activated protein kinase (AMPK) restoring the cell energy or ATP. Both mechanisms have been associated with SDS consumption since in vivo studies show the activation of the AMPK with SDS intake [13, 62, 63].

Other authors suggest that the benefits in the glucose homeostasis derived from the RS intake are a consequence of improved lipid homeostasis. The improvement in lipid metabolism is modulated through the promotion of muscular lipid oxidation and cholesterol homeostasis, both related to the improvements in glucose homeostasis. Whereas, other authors suggest that the improvement of the glucose homeostasis is a consequence of the SCFA released, contributing to promote the satiety process and improving the serum lipid profile, both related to an improvement in the prognostic of the insulin resistance [56, 58, 59, 61].

Many researchers suggest that SDS or RS consumption improves the glycaemia in diabetics through the activity reduction of the enzymes of the hepatic gluconeogenesis and the augmentation of the glucose intake by the muscle tissue [58, 59, 61], as well as by the improvement of the lipid homeostasis consequence of the release of SCFA during colonic fermentation of RS [26, 53, 57]. The improvement of glucose homeostasis is an efficient mechanism to reduce the long-term complications consequence of diabetes, mainly related to oxidative stress and cell damage [61–63]. In this context, the glycemic response in healthy individuals and type 2 diabetic patients is correlated with the type and amount of RS ingested [63, 64]. Other biomarkers related to glucose modulation are the C peptide, leptin, PYY, GLP-1, GIP, and some inflammatory cytokines. The RS consumption is associated with the increase of plasmatic levels of GLP-1, GIP, and PYY, responsible for the insulinemic response, glucose regulation, and satiety process. The released SCFA during RS colonic fermentation could be associated with the increase in the expression of PYY and GLP-1 genes, establishing a relation between RS consumption and intestinal hormones production [63, 64]. Maziarz et al. [65] demonstrated that RS consumption reduces leptin production probably a consequence of augmentation on the fatty acids oxidation since serum circulating levels of leptin are associated with the total body fat mass and negatively correlated with fatty acids oxidation.

5. SDS and RS as potential functional ingredients

As a consequence of the improved awareness of the consumers in the relation of diet and health, the food industry has focused on the production of functional foods based on cereals, legumes, and other products with low glycemic index (GI) [11–13, 25, 29, 32]. Currently, new sources for RS obtention or production have been investigated as alternatives to the conventional sources (maize, wheat, rice, among others) since RS beneficial effects are dependent on the botanical origin, type of RS, and total intake amount [12, 25, 29, 32]. RS is shown as a great potential functional ingredient because of their great techno-functional properties such as small particle size, color, soft flavor, well properties for the extrusion process, high temperatures of gelatinization, and low water retention capacity, together with their low caloric values (1. 6–2.8 kcal/g) [5, 11, 25]. In addition, may improve the lifespan of dry products and avoid the ice crystals during the ice creams production [5, 25, 29, 46]. These properties allow to incorporate RS as an ingredient in many different food matrixes such as dairy products, bakery products, or pastas [5, 11, 25, 46].

RS can be used also to increase the total fiber content of food products or to reduce the low available CH contents in the production of dietetic food products focused on reducing the bodyweight through a lower caloric content [25, 46]. The incorporation of RS to the food formulation does not change the flavor neither produce significant changes in the texture, whereas may improve the product sensorial properties compared with many available fibers such as brans or gums [11, 25, 46], also, the final product texture can be improved by using RS because of their low water retention capacity [29, 46–48]. RS has been also used as a cover ingredient during probiotics microencapsulation to be incorporated into dairy products, allowing increasing the viability of the probiotics. Also, RS has been used to encapsulate fish oils to reduce the odor and lipid oxidation [5, 25, 46, 47]. RS can be used in bakery products and breakfast cereals as a functional ingredient and fiber source but attention to the technological properties of the products should be observed to ensure to reach the desirable properties on the product [5, 11, 25, 29, 46]. The addition of RS-III to sourdough breaths may improve the toasted color and the crunch properties of the product [25, 46, 47]. However, the addition of green banana flour (rich in RS-II) to pasta may provoke a weaker mass as a consequence of the gluten dilution effect but less oil is required to fry these products [5, 25, 46]. The addition of RS-III ha can generate in the flour tortillas lower flexibility, rolling capacity, and cohesion. High concentrations of RS-III reduce structural integrity and therefore product quality [5, 25, 29, 32, 46].

On the other hand, the evaluation of the impact and benefits in the consumption of SDS as a functional ingredient for the elaboration of products that improve the postprandial glycemic response has increased in recent years. Rebello et al. [66] evaluated the effect of the consumption of SDS as a functional ingredient in a Snack, with the purpose of improving the postprandial response, the degree of satisfaction, and appetite in people with overweight and obesity. The consumption of the snack with SDS at breakfast improved the postprandial glycemic and insulinemic response by decreasing the concentrations of glucose and serum insulin in the blood per unit of time (minutes), compared to a control product without SDS. In addition, they estimated that the relative glucose responses in the evaluated products were 40% lower compared to the control product. They attribute this effect to the characteristics and type of CH ingested, and to a lower rate of digestion and absorption of the same. These results are consistent with that reported by Péronnet et al. [67], where they evaluated the consumption of a cereal product with different SDS contents (high and low) to evaluate the type of glycemic response in relation to the underlying changes in the kinetics of plasma glucose, on the increase in the rate of production (appearance) and consumption (disappearance) of serum glucose in the blood in healthy young women. They observed that the consumption of the product with a high content of SDS generated a significant reduction in the absorption and consumption kinetics of plasma glucose concentrations, compared to the consumption of the control products. This group considers that the consumption of the product with a high content of SDS generates a slow absorption of glucose, reducing its availability, and promotes a continuous gradual effect, attenuating sudden changes in concentration in the kinetics of plasma glucose, improving its homeostasis.

6. Conclusions

The impact of starch consumption on the human health is dependent on the starch composition and physicochemical properties, highly dependent on the botanical origin of the starch or the processes applied. In this context, the major factor associated with the impact of starch consumption on the human health is the hydrolysis behavior during the gastrointestinal tract, and therefore, with the composition and amount of RDS, SDS, and RS fraction in the starch molecule. Starches with high RDS amounts are characterized by rapid and high glycaemic and insulinemic peaks, being a risk for further diabetes and other chronic diseases. Conversely, SDS and especially RS are associated with low or very low but maintained glycemic and insulinemic responses, and therefore, are associated with benefits for the glucose homeostasis in diabetics but also in healthy people. RS is shown as a promising functional ingredient since this molecule, depending on its morphological structure and physicochemical characteristics, is low or non-digested and shows other potentially beneficial properties. The majority of RS is fermented by the colonic microbiota allowing to growth of the Bifidobacterias and Lactobacillus species, acting as a prebiotic compound. Also, SCFA is released from their colonic fermentation, associated with the increase in the expression of satiety hormones such as GLP-1, GIP, and PYY, contributing to reducing the total calories intake. It has been suggested that RS and SDS consumption is also associated with the glucose homeostasis, contributing to reduce the expression of the gluconeogenic enzymes (PEPCK and G6Pase) and improving the glucose intake by the muscle tissue. In addition, RS and SDS may contribute to lipid homeostasis, increasing HDL release and lipid oxidation. However, more detailed studies are required to clarify the capacity of RS and SDS to modulate glucose and lipid homeostasis. In conclusion, starches with high SDS and especially RS are associated with health benefits such as low insulinemic responses, glucose homeostasis control, prebiotic effects, and satiety, being RS as a promising but low exploited functional ingredient.

Conflict of interest

The authors declare that they have no known competing financial interests or personal relationships that could have appeared to influence the work reported in this paper.

Author details

Teodoro Suarez-Diéguez[1*] and Juan Antonio Nieto[2,3]

1 Academic Area of Nutrition, Institute of Health Sciences, Autonomous University of the State of Hidalgo, Pachuca de Soto, Hidalgo, México

2 AINIA Technological Centre, Paterna, Valencia, Spain

3 International University of Valencia, Valencia, Spain

*Address all correspondence to: tsuarez@uaeh.edu.mx

IntechOpen

© 2021 The Author(s). Licensee IntechOpen. This chapter is distributed under the terms of the Creative Commons Attribution License (http://creativecommons.org/licenses/by/3.0), which permits unrestricted use, distribution, and reproduction in any medium, provided the original work is properly cited.

References

[1] Gul K, Singh AK, Jabeen R. Nutraceuticals and functional foods: The foods for the future world. Critical Reviews in Food Science and Nutrition. 2016;**56**(16):2617-2627. DOI: 10.1080/10408398.2014.903384

[2] Galanakis CM. Introduction. In: Galanakis CM, editor. Nutraceutical and Functional Food Components. Academic Press is an imprint of Elsevier; 2017. pp. 1-14. DOI: 10.1016/B978-0-12-805257-0.00001-6

[3] Gerschenson LN, Rojas AM, Fissore EN. Carbohydrates. In: Galanakis CM, editor. Nutraceutical and Functional Food Components. Academic Press is an imprint of Elsevier; 2017. pp. 39-101. DOI: 10.1016/B978-0-12-805257-0.00003-X

[4] Lovegrove A, Edwards CH, De Noni I, Patel H, El SN, Grassby T, et al. Role of polysaccharides in food, digestion, and health. Critical Reviews in Food Science and Nutrition. 2017;**57**(2):237-253. DOI: 10.1080/10408398.2014.939263

[5] Sharma A, Sing-Yadav B, Ritika B. Resistant starch: Physiological roles and food applications. Food Review International. 2008;**24**(2):193-234. DOI: 10.1080/87559120801926237

[6] Englyst KN, Englyst HN. Carbohydrate bioavailability. The British Journal of Nutrition. 2005;**94**(1):1-11. DOI: 10.1079/BJN20051457

[7] Singh J, Dartois A, Kaur L. Starch digestibility in food matrix: A review. Trends in Food Science and Technology. 2010;**21**(4):168-180. DOI: 10.1016/j.tifs.2009.12.001

[8] Butterworth PJ, Warren FJ, Ellis PR. Human α-amylase and starch digestion: An interesting marriage. Starch/Stärke. 2011;**63**(7):395-405. DOI: 10.1002/star.201000150

[9] Lapis TJ, Penner MH, Balto AS, Lim J. Oral digestion and perception of starch: Effects of cooking, tasting time, and salivary α-amylase activity. Chemical Senses. 2017;**42**(8):635-645. DOI: 10.1093/chemse/bjx042

[10] Hoover R, Zhou Y. In vitro and in vivo hydrolysis of legume starches by α-amylase and resistant starch formation in legumes—a review. Carbohydrate Polymers. 2003;**54**(4):401-417. DOI: 10.1016/S0144-8617(03)00180-2

[11] Fuentes-Zaragoza E, Riquelme-Navarrete MJ, Sánchez-Zapata E, Pérez-Álvarez JA. Resistant starch as functional ingredient: A review. Food Research International. 2010;**43**(4):931-942. DOI: 10.1016/j.foodres.2010.02.004

[12] Lockyer S, Nugent AP. Health effects of resistant starch. Nutrition Bulletin. 2017;**42**(1):10-14. DOI: 10.1111/nbu.12244

[13] Ting-Wong TH, Yu-Louie JC. The relationship between resistant starch and glycemic control: A review on current evidence and possible mechanisms. Starch/Stärke. 2017;**69**(7-8):1600205. DOI: 10.1002/star.201600205

[14] Nazare JA, de Rougemont A, Normand S, Sauvinet V, Sothier M, Vinoy S, et al. Effect of postprandial modulation of glucose availability: Short- and long-term analysis. The British Journal of Nutrition. 2010;**103**(10):1461-1470. DOI: 10.1017/S0007114509993357

[15] Vinoy S, Meynier A, Goux A, Jourdan-Salloum N, Normand S,

Rabasa-Lhoret R, et al. The effect of a breakfast rich in slowly digestible starch on glucose metabolism: A statistical meta-analysis of randomized controlled trials. Nutrients. 2017;**9**(4):318. DOI: 10.3390/nu9040318

[16] Dhital S, Warren FJ, Butterworth PJ, Ellis PR, Gidley MJ. Mechanisms of starch digestion by α-amylase: Structural basis for kinetic properties. Critical Reviews in Food Science and Nutrition. 2017;**57**:875-892. DOI: 10.1080/10408398.2014.922043

[17] Edwards CH, Warren FJ, Milligan PJ, Butterworth PJ, Ellis PR. A novel method for classifying starch digestion by modelling the amylolysis of plant foods using first-order enzyme kinetic principles. Food & Function. 2014;**5**(11):2751-2758. DOI: 10.1039/C4FO00115J

[18] Butterworth PJ, Warren FJ, Grassby T, Patel H, Ellis PR. Analysis of starch amylolysis using plots for first-order kinetics. Carbohydrate Polymers. 2012;**87**(3):2189-2197. DOI: 10.1016/j.carbpol.2011.10.048

[19] Gourineni V, Stewart ML, Skorge R, Sekula BC. Slowly digestible carbohydrate for balanced energy: In vitro and in vivo evidence. Nutrients. 2017;**9**(11):1230. DOI: 10.3390/nu9111230

[20] Marinangeli CP, Harding SV. Health claims using the term 'sustained energy' are trending but glycaemic response data are being used to support: Is this misleading without context? Journal of Human Nutrition and Dietetics. 2016;**29**(4):401-404. DOI: 10.1111/jhn.12359

[21] Van-Dam RM, Seidell JC. Carbohydrate intake and obesity. European Journal of Clinical Nutrition. 2007;**61**(Suppl. 1):S75-S99. DOI: 10.1038/sj.ejcn.1602939

[22] Goodman BE. Insights into digestion and absorption of major nutrients in humans. Advances in Physiology Education. 2010;**34**(2):44-53. DOI: 10.1152/advan.00094.2009

[23] Martens BM, Bruininx EM, Gerrits WJ, Schols HA. The importance of amylase action in the porcine stomach to starch digestion kinetics. Animal Feed Science and Technology. 2020;**267**:114546

[24] Cornejo-Ramírez YI, Martínez-Cruz O, Toro-Sánchez CLD, Wong-Corral FJ, Borboa-Flores J, Cinco-Moroyoqui FJ. The structural characteristics of starches and their functional properties. CyTA—Journal of Food. 2018;**16**(1):1003-1017. DOI: 10.1080/19476337.2018.1518343

[25] Sofi SA, Ayoub A, Jan A. Resistant starch as functional ingredient: A review. International Journal of Food Sciences and Nutrition. 2017;**2**(6):195-199. DOI: 10.1016/j.foodres.2010.02.004

[26] Keenan MJ, Zhou J, Hegsted M, Pelkman C, Durham HA, Coulon DB, et al. Role of resistant starch in improving gut health, adiposity, and insulin resistance. Advances in Nutrition. 2015;**6**(2):198-205. DOI: 10.3945/an.114.007419

[27] Ma M, Wang Y, Wang M, Jane J-l, Du S-k. Physicochemical properties and in vitro digestibility of legume Starches. Food Hydrocolloids. 2017;**63**:249-255. DOI: 10.1016/j.foodhyd.2016.09.004

[28] Zhang B, Li H, Wang S, Junejo SA, Liu X, Huang Q. In vitro starch digestion: Mechanisms and kinetic models. In: Wang S, editor. Starch Structure, Functionality and Application in Foods.

Singapore: Springer Nature; 2020. pp. 151-167. DOI: 10.1007/978-981-15-0622-2_9

[29] Ren Y, Yuan TZ, Chigwedere CM, Ai Y. A current review of structure, functional properties, and industrial applications of pulse starches for value-added utilization. Comprehensive Reviews in Food Science and Food Safety. 2021;**20**:3061-3092. DOI: 10.1111/1541-4337.12735

[30] Dona AC, Pages G, Gilbert RG, Kuchel PW. Digestion of starch: In vivo and in vitro kinetic models used to characterize oligosaccharide or glucose release. Carbohydrate Polymers. 2010;**80**(3):599-617. DOI: 10.1016/j.carres.2017.05.016

[31] Li Y, Xu J, Zhang L, Ding Z, Gu Z, Shi G. Investigation of debranching pattern of a thermostable isoamylase and its application for the production of resistant starch. Carbohydrate Research. 2017;**446-447**:93-100. DOI: 10.1016/j.carres.2017.05.016

[32] Suárez-Diéguez T, Pérez-Moreno F, Ariza-Ortega JA, López-Rodríguez G, Nieto JA. Obtention and characterization of resistant starch from creole faba bean (*Vicia faba* L. creole) as a promising functional ingredient. LWT. 2021;**145**:111247. DOI: 10.1016/j.lwt.2021.111247

[33] Yu W, Tao K, Gilbert RG. Improved methodology for analyzing relations between starch digestion kinetics and molecular structure. Food Chemistry. 2018;**264**:284-292. DOI: 10.1016/j.foodchem.2018.05.049

[34] Guo L, Xiao Y, Zhu C, Wang S, Du X, Cui B. In vitro enzymatic hydrolysis of amylopectins from rice starches. International Journal of Biological Macromolecules. 2017;**105**:1001-1009. DOI: 10.1016/j.ijbiomac.2017.07.138

[35] Bertoft E. Understanding starch structure: Recent progress. Agronomy. 2017;**7**(3):56. DOI: 10.3390/agronomy7030056

[36] Ai Y, Jane J-l. Understanding starch structure and functionality. In: Sjöö M, Nilsson L, editors. Starch in Food. 2nd ed. Ámsterdam: Woodhead Publishing Series in Food Science, Technology and Nutrition; 2018. pp. 151-178. DOI: 10.1016/B978-0-08-100868-3.00003-2

[37] Robyt JF. Enzymes and their action on starch. In: Be Miller J, Whistler R, editors. Starch, Chemistry and Technology Food Science and Technology. 3rd ed. London, United Kingdom: Academic Press; 2009. pp. 237-292. DOI: 10.1016/B978-0-12-746275-2.00007-0

[38] Ambigaipalan P, Hoover R, Donner E, Liu Q, Jaiswal S, Chibbar R, et al. Structure of faba bean, black bean and pinto bean starches at different levels of granule organization and their physicochemical properties. Food Research International. 2011;**44**(9):2962-2974. DOI: 10.1016/j.foodres.2011.07.006

[39] Dhital S, Shrestha AK, Gidley MJ. Relationship between granule size and in vitro digestibility of maize and potato starches. Carbohydrate Polymers. 2010;**82**(2):480-488. DOI: 10.1016/j.carbpol.2010.05.018

[40] Villas-Boas F, Facchinatto WM, Colnago LA, Volanti DP, Franco CML. Effect of amylolysis on the formation, the molecular, crystalline and thermal characteristics and the digestibility of retrograded starches. International Journal of Biological Macromolecules. 2020;**163**:1333-1343. DOI: 10.1016/j.ijbiomac.2020.07.181

[41] Walther B, Lett AM, Bordoni A, Tomás-Cobos L, Nieto JA, Dupont D, et al. GutSelf: Interindividual variability in the processing of dietary compounds by the human gastrointestinal tract. Molecular Nutrition & Food Research. 2019;**63**(21):1900677. DOI: 10.1002/mnfr.201900677

[42] Datc K. Regulatory functions of α-amylase in the small intestine other than starch digestion: α-Glucosidase activity, glucose absorption, cell proliferation, and differentiation. In: Takada A, editor. Metabolic Syndrome. London: IntechOpen; 2020. pp. 1-14. DOI: 10.5772/intechopen.92660

[43] Englyst KN, Liu S, Englyst HN. Nutritional characterization and measurement of dietary carbohydrates. European Journal of Clinical Nutrition. 2007;**61**(Suppl. 1):S19-S39. DOI: 10.1038/sj.ejcn.1602937

[44] Zhang G, Hamaker BR. Slowly digestible starch: Concept, mechanism, and proposed extended glycemic index. Critical Reviews in Food Science and Nutrition. 2009;**49**(10):852-867. DOI: 10.1080/10408390903372466

[45] Toutounji MR, Farahnaky A, Santhakumar AB, Oli P, Butardo VM, Blanchard CL. Intrinsic and extrinsic factors affecting rice starch digestibility. Trends in Food Science and Technology. 2019;**88**:10-22. DOI: 10.1016/j.tifs.2019.02.012

[46] Raigond P, Ezekiel R, Raigond B. Resistant starch in food: A review. Journal of the Science of Food and Agriculture. 2015;**95**(10):968-1978. DOI: 10.1002/jsfa.6966

[47] Ma Z, Boye JI. Research advances on structural characterization of resistant starch and its structure-physiological function relationship: A review. Critical Reviews in Food Science and Nutrition. 2018;**58**(7):1059-1083. DOI: 10.1080/10408398.2016.1230537

[48] Ma Z, Hu X, Boye JI. Research advances on the formation mechanism of resistant starch type III: A review. Critical Reviews in Food Science and Nutrition. 2020;**60**(2):276-297. DOI: 10.1080/10408398.2018.1523785

[49] Frank J, Fukagawa NK, Bilia AR, Johnson EJ, Kwon O, Prakash V, et al. Terms and nomenclature used for plant-derived components in nutrition and related research: Efforts toward harmonization. Nutrition Reviews. 2020;**78**(6):451-458. DOI: 10.1093/nutrit/nuz081

[50] Breyton AE, Goux A, Lambert-Porcheron S, Meynier A, Sothier M, VanDenBerghe L, et al. Starch digestibility modulation significantly improves glycemic variability in type 2 diabetic subjects: A pilot study. Nutrition, Metabolism, and Cardiovascular Diseases. 2021;**31**(1):237-246. DOI: 10.1016/j.numecd.2020.08.010

[51] Caprnda M, Mesarosova D, Ortega PF, Krahulec B, Egom E, Rodrigo L, et al. Glycemic variability and vascular complications in patients with type 2 diabetes mellitus. Folia Medica (Plovdiv). 2017;**59**(3):270-278. DOI: 10.1515/folmed-2017-0048

[52] Sakamoto M. Type 2 diabetes and glycemic variability: Various parameters in clinical practice. Journal of Clinical Medical Research. 2018;**10**(10):737-742. DOI: 10.14740/jocmr3556w

[53] Hasek LY, Phillips RJ, Zhang G, Kinzig KP, Kim CY, Powley TL, et al. Dietary slowly digestible starch triggers the gut-brain axis in obese rats with accompanied reduced food intake. Molecular Nutrition & Food Research.

2018;**62**(5):1700117. DOI: 10.1002/mnfr.201700117

[54] Lambert-Porcheron S, Normand S, Blond E, Sothier M, Roth H, Meynier A, et al. Modulation of starch digestibility in breakfast cereals consumed by subjects with metabolic risk: Impact on markers of oxidative stress and inflammation during fasting and the postprandial period. Molecular Nutrition & Food Research. 2017;**61**(12):1700212. DOI: 10.1002/mnfr.201700212

[55] Khosroshahi HT, Abedi B, Ghojazadeh M, Samadi A, Jouyban A. Effects of fermentable high fiber diet supplementation on gut derived and conventional nitrogenous product in patients on maintenance hemodialysis: A randomized controlled trial. Nutrition & Metabolism (London). 2019;**16**:18. DOI: 10.1186/s12986-019-0343-x

[56] Halajzadeh J, Milajerdi A, Reiner Ž, Amirani E, Kolahdooz F, Barekat M, et al. Effects of resistant starch on glycemic control, serum lipoproteins and systemic inflammation in patients with metabolic syndrome and related disorders: A systematic review and meta-analysis of randomized controlled clinical trials. Critical Reviews in Food Science and Nutrition. 2020;**60**(18):3172-3184. DOI: 10.1080/10408398.2019.1680950

[57] Cani PD, Delzenne NM. The role of the gut microbiota in energy metabolism and metabolic disease. Current Pharmaceutical Design. 2009;**15**(13):1546-1558. DOI: 10.2174/138161209788168164

[58] Sun H, Ma X, Zhang S, Zhao D, Liu X. Resistant starch produces antidiabetic effects by enhancing glucose metabolism and ameliorating pancreatic dysfunction in type 2 diabetic rats. International Journal of Biological Macromolecules. 2018;**110**:276-284. DOI: 10.1016/j.ijbiomac.2017.11.162

[59] Zhou Z, Wang F, Ren X, Wang Y, Blanchard C. Resistant starch manipulated hyperglycemia/hyperlipidemia and related genes expression in diabetic rats. International Journal of Biological Macromolecules. 2015;**75**:316-321. DOI: 10.1016/j.ijbiomac.2015.01.052

[60] Oikonomou E, Tsioufis C, Tousoulis D. Diabetes mellitus: A primary metabolic disturbance. Metabolomics underlying vascular responses to stress and ischemia? Clinical Science. 2020;**135**(3):589-591. DOI: 10.1042/CS20201299

[61] Hatting M, Tavares C, Sharabi K, Rines AK, Puigserver P. Insulin regulation of gluconeogenesis. Annals of the New York Academy of Sciences. 2018;**1411**(1):21-35. DOI: 10.1111/nyas.13435

[62] Dong Y, Gao G, Fan H, Li S, Li X, Liu W. Activation of the liver X receptor by agonist TO901317 improves hepatic insulin resistance via suppressing reactive oxygen species and JNK pathway. PLoS One. 2015;**10**(4): e0124778. DOI: 10.1371/journal.pone.0124778

[63] Nuttall FQ, Ngo A, Gannon MC. Regulation of hepatic glucose production and the role of gluconeogenesis in humans: Is the rate of gluconeogenesis constant? Diabetes/Metabolism Research and Reviews. 2008;**24**(6):438-458. DOI: 10.1080/10408398.2020.1747391

[64] Guo J, Tan L, Kong L. Impact of dietary intake of resistant starch on obesity and associated metabolic profiles in human: A systematic review of the literature. Critical Reviews in Food Science and Nutrition. 2021;**61**(6):

889-905. DOI: 10.1080/10408398.2020.1747391

[65] Maziarz MP, Preisendanz S, Juma S, Imrhan V, Prasad C, Vijayagopal P. Resistant starch lowers postprandial glucose and leptin in overweight adults consuming a moderate-to-high-fat diet: A randomized-controlled trial. Nutrition Journal. 2017;**16**(1):14. DOI: 10.1186/s12937-017-0235-8

[66] Rebello CJ, Johnson WD, Pan Y, Larrivee S, Zhang D, Nisbet M, et al. A snack formulated with ingredients to slow carbohydrate digestion and absorption reduces the glycemic response in humans: A randomized controlled trial. Journal of Medicinal Food. 2020;**23**(1):21-28. DOI: 10.1089/jmf.2019.0097

[67] Péronnet F, Meynier A, Sauvinet V, Normand S, Bourdon E, Mignault D, et al. Plasma glucose kinetics and response of insulin and GIP following a cereal breakfast in female subjects: Effect of starch digestibility. European Journal of Clinical Nutrition. 2015;**69**(6):740-745. DOI: 10.1038/ejcn.2015.50

Chapter 5

Potato Starch as Affected by Varieties, Storage Treatments and Conditions of Tubers

Saleem Siddiqui, Naseer Ahmed and Neeraj Phogat

Abstract

Potato is among the widely grown crop of the world. It is likely that a large portion of the crop is consumed fresh but majority of it is processed into various products, starch being the predominant one. Starch can greatly contribute to the textural properties of many foods and is widely used in food industry as raw material. Since raw potatoes are perishable and accessible only for few months of the year, the food and starch industry has to rely on stored potatoes during off-season. The various varieties of the crop available in the region, storage conditions, pre and post-storage treatments given to the tubers, packaging materials used, etc. are influencing the physical, chemical and functional characteristics of starch extracted from it. The extraction technology from tubers is also having a significant effect on the quality of starch. The knowledge of physical, chemical and functional characteristics of potato starch as affected by varieties, storage treatments and conditions of tubers will help in ensuring uniform and desirable quality of starch for food industry and also provide information for breeding programs and developing the proper postharvest management practices of potatoes.

Keywords: curing, extraction methods, functional properties, packaging potato starch, sprout suppressants, storage conditions, varieties

1. Introduction

Potato (*Solanum tuberosum* L.) is the most important food crop in the world after wheat, rice, and maize. UNESCO (United Nations Educational, Scientific and Cultural Organization) declared potato as the food of the future during the 'International Year of Potato 2008' and stated potato as the third most important world food crop. Potato production increased significantly in India during the last six decades and it became the second-largest producer of potatoes after China [1]. Potatoes contain 70–80% water, 16–24% starch (85–87% dry mass), and trace amounts of proteins and lipids [2].

Potatoes are a perishable crop, and due to insufficient, expensive, and widely dispersed refrigerated storage facilities, there are frequent instances of market over-supply, resulting in significant economic damage to farmers and agricultural wastage.

Various storage treatments and technologies have been proposed to extend the shelf life of potato tubers. Because raw potatoes are accessible only for few months of the year, the food and starch industry has to rely on stored potatoes during off-season. A proper storage climate helps keep potatoes in excellent condition by avoiding excessive weight loss, microbial spoilage, sprout development, and quality degradation. Potatoes are often stored in long-term postharvest cold storage (8–12°C, 85–90% RH) to retard physiological processes and extend shelf life. Maintaining low temperatures during the storage time is dependent on the tubers' intended usage

Prolonged storage of potatoes requires sprout inhibition either by use of irradiation or sprout inhibitor chlorpropham (CIPC, isopropyl 3-chlorocarbanilate) treatment [3, 4] or the usage of heat treatment [5]. The CIPC alternatives for sprout inhibition are maleic hydrazide (MH), 1,4-Dimethylnaphthalene, and ethylene [6]. Heat treatment, essential oils of some herbs and spices—also effectively reduce sprouting and can be applied to organically grown potatoes [7, 8]. The sprout inhibiting treatments besides affecting the physiology of tubers also alter the properties of starch [9].

Starch plays a major role in the sensory characteristics of a wide variety of foods and is extensively used in agro-industrial applications as a thickener, colloidal stabilizer, gelling agent, bulking agent, and water retention agent. Because of the higher granule size and purity of potato starch, as well as the amylose and amylopectin chain lengths and the presence of phosphate ester groups on amylopectin, it stands apart from other cereal starches (corn, wheat, rice, etc.). Potato starch is an excellent texture stabiliser and regulator in food production systems.

The adaptability of starch in a wide range of food items makes it a hot topic among researchers studying carbohydrate polymers. As the genetic basis of the starch changes, so do its physicochemical attributes and functional features, as well as how unique they are in different foods and drinks [10]. Even within the same botanical species, cultivars of the same plant grown in different environments and cultural settings have vastly different starch structures and functions. This diversity results in a wide variety of starches with varying cooking, textural, and rheological characteristics that are linked to their physicochemical, morphological, and thermal properties. The following sections of the chapter deal with the starch characteristics as influenced by extraction methods, cultivars, curing, sprout inhibitors and storage conditions of potato tubers.

2. Effect of extraction methods

Depending on the plant source and intended application for the starch, several techniques can be used to extract it. Different researchers have utilized a variety of extraction methods to separate starch, including steeping periods, extraction temperatures, chemical concentrations and nature, enzymes, and so on. The chemical composition and physical characteristics of starch are both influenced by the extraction processes. To procure a pure product with maximum yield and recovery, lowest cost, and application of a series of interrelated stages allowing the non-starchy fraction to be removed without affecting the granule native structure and minimal incidence on its physico-chemical and mechanical properties, selecting an appropriate starch extraction method is desirable.

The chemical and physical characteristics of starch are altered throughout the extraction process. According to Neeraj *et al.* [11], water temperature affected potato tuber starch extraction. Extraction temperatures of 30°C and 60°C produced improved starch yields, water absorption capacity (WAC), and whiteness. The use of 0.25% NaOH for chemical extraction increased WAC while decreasing ash content. The combination of sodium dodecyl sulphate (SDS) and mercaptoethanol (ME) to remove lipids increased amylose and starch purity while reducing amylopectin, moisture, and fat. Whiteness values were greater in the extraction employing NaOCl bleaching agent, while starch yields were higher in the enzymatic extraction by cellulose. The combination treatment of NaOH, SDS-ME, Na-hypochlorite and cellulase produced significantly higher starch yield and WAC than the simple cold water extraction method. It also resulted in significantly higher swelling power and solubility, light transmittance and whiteness, as well as trough and final viscosity of the extracted starch.

The different extraction techniques resulted in varying proportions of tiny and big starch granules. The Na-bisulphite treatment had the highest percentage of small-sized particles, whereas the other methods had no noticeable variations in particle size. Intact starch granules with smooth surfaces were found in the water-treated starch, but granules of starch treated with Na-bisulphite or propanol-water had somewhat rough and pitted surfaces, as well as fractures inside the granules (**Figure 1**) [11]. Lin *et al.* [12] reported that alcohol treatment caused not only the disappearance of 'Maltese cross' pattern in the center of granule but also the occurrence of cracks inside the granule. Betancur *et al.* [13] observed that Na-bisulphite treatment resulted in acid hydrolysis, with the hydroxonium ion attacking the glycosidic oxygen atom and hydrolyzing the glycosidic linkage, altering the physicochemical properties of starch and causing the formation of some cracks on the starch surface.

Phosphorus is present as phosphate monoesters and phospholipids in potato starches [14]. Phosphorus alters the functional characteristics of starches, which is significant both technologically and nutritionally. The phosphate monoesters are covalently linked to the starch's amylopectin portion. The majority of the phosphate groups are covalently linked to the amylopectin fraction at the C-6 (70%) and C-3 (30%) locations of the glucose units. It has been demonstrated that the phosphate concentration of starches has an effect on their physicochemical qualities and end applications, including starch pasting capabilities, gel strength and clarity, stickiness and viscosity [15]. Neeraj *et al.* [11] reported that starch extracted by water had higher, while extracted by Na-hypochlorite lower phosphorus content. The decrease in phosphorus content by Na-hypochlorite (NaOCl) was attributed to acid hydrolysis of amylose and amylopectin chains, reducing the ash content of starch and hence its mineral content. NaOCl may have caused oxidative degradation of amylopectin and amylose chain and thereby reducing the phosphorus content [16].

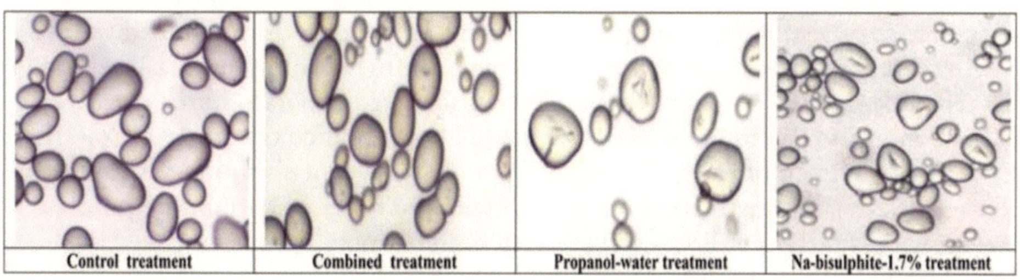

Figure 1.
Effect of various methods of extraction on potato starch granule (source: [11]).

The variations in WAC between different starches are attributable to the degree to which water binding sites are available in their granules. The hydroxyl groups and inter-glucose oxygen atoms are thought to be the binding sites. The capacity of these starches to interact with water is determined by their ultrastructural (molecular arrangement, amorphous and crystalline regions) and compositional variations (primarily amylose and amylopectin). The capacity of commercial starches to bind water is critical to the quality and texture of some food items because it protects them from effects such as syneresis, which can occur during retorting or freezing [15]. It has been reported that different extraction techniques significantly affected WAC of potato starch. The alkali treatment with NaOH resulted in a greater WAC, but the NaOCl extraction technique resulted in a lower WAC [11, 17]. The increased WAC in alkali treated starch might be due to ions (Na^+) diffusing into the amylose-rich amorphous areas of the granules, destroying intermolecular interactions, altering the starch's crystalline structure, and leading the granules to absorb more water [18, 19]. Fat and protein that are located on the surface of starch granules are crucial for maintaining its structural stability [20]. Additionally, their presence has been demonstrated to significantly slow the rate of starch retrogradation [21]. Lipids have also been observed to form complexes with amylopectin's outer branches, therefore inhibiting starch retrogradation. It has been reported by Neeraj *et al* [11] that different extraction techniques have an effect on the protein and fat content of potato starch. The combined treatment of NaOH, SDS-ME, Na-hypochlorite and cellulose resulted in low protein content, but the cold water extracted potato starch had high protein content. The decreased protein content seen in combination treatment may be a result of the alkali, SDS, and protease present. Because NaOH is an excellent solvent and can solubilize the main protein enclosing the starch, alkaline steeping technique softens the protein-starch matrix, resulting in a starch deficient in protein and lipids [22]. SDS treatment was also found to be effective in removing the protein and lipids from the surface of starch granules [20].

Different extraction treatments resulted in significant variations in swelling power and solubility index of the extracted starch. Neeraj *et al.* [11] reported that the swelling power and solubility were maximum for starch extracted by NaOH, while it was minimum in Na-bisulphite. Wang and Wang [23] reported that NaOH treatment caused the removal of protein and lipids from the surface of the starch granules and then allowed the starch granules to swell more and open up the small pores or crevices on the granule surface. Sajeev and Moorthy [24] reported that reduced swelling power of sulphite-treated starches may be owing to the fact that sulphite interacts stoichiometrically with oxygen (2 moles of sulphite to 1 mole of oxygen), forming sulphite free radicals and the superoxide ion ($^{\cdot}O^{2}$-).

The freeze-thaw stability (syneresis) of starch is a helpful indication of its retrograde tendency. Syneresis is a critical characteristic that is used to determine a starch's capacity to survive unfavourable physical changes that occur during freezing and thawing. Extraction procedures also have a major influence on syneresis. Neeraj *et al* [17] reported that the maximum synerisis was observed in potato starch extracted by Na-bisulphite and minimum in starch extracted by combined treatment of NaOH, SDS-ME, Na-hypochlorite and cellulose. The enhanced syneresis induced by starch following Na-bisulphite treatment may be attributed to the acid-thinning process increasing the fraction of linear chain starch in the sample, which enhanced the inclination to retrograde.

Granule swelling, granule remnants, leached amylose and amylopectin, as well as the molecular weights and chain lengths of amylose and amylopectin, have all been reported to vary with granule size, resulting in the development of turbidity and reduced light transmittance in refrigerated starch pastes [25]. It has been reported by Neeraj *et al.* [11] that the potato starch extracted by Na-hypochlorite treatment showed higher transmittance than native starch because the oxidized starch had a lower tendency for molecular re-association. The presence of hydrophilic functional groups in oxidised starches, particularly carboxyl groups, may account for the increased transmittance. It can also be ascribed to the chemical replacement of carbonyl and carboxyl functional groups for the hydroxyl groups in starch molecules, leading to repulsion between neighbouring starch molecules and decreased inter-chain interaction, allowing for increased transmittance [26].

3. Effect of potato varieties on starch

Tuber starches from different potato breeds vary in terms of crystallinity, granule shape, and other physical and chemical characteristics. The reactivity of the starches in various potato genotypes varied significantly [27]. The accumulation of starch in potatoes is genotype- and environment-dependent, as well as genotype-environment interaction dependent [28]. Due to the variations in tuber development rates amongst cultivars, the harvest dates and hence the dry mater accumulation varied for different potato cultivars. Early maturing cultivars showed lower dry matter content and a lower starch content than late maturing types. Kufri Chipsona-4 produced the most, followed by Kufri Badshah, Kufri Sindhuri, and Kufri Bahar; and Kufri Pushkar produced the least amount of starch. Since Kufri Chipsona-4 is a medium to late maturing variety, it produced more starch than other cultivars [11].

Kaur *et al.* [29] screened 21 different potato varieties and reported lowest amylose content of 15.0% for Kufri Ashoka (Patna) and the highest of 23.1% for Kufri Badshah (Jalandhar) starch. Singh *et al.* [30] compared the starch amylose content among the potato varieties and reported that Kufri Jyoti starch had the highest amylose content, whereas Kufri Sindhuri starch had the lowest. The variations in amylose concentration across types of starch granules have been attributed to the activity of enzymes involved in the production of linear and branching components inside the starch granules during plant development [11].

Singh and Singh [31] reported that Kufri Badshah (KB) and Kufri Jyoti (KJ) starch paste showed higher light transmittance and lower turbidity values than Kufri Pukhraj (KP) potato starch pastes. Kaur *et al.* [32] observed that starches separated from varieties KJ and KB had lower transition temperatures (To; Tp and Tc), peak height indices (PHI), higher gelatinization temperature range (R) and enthalpies of gelatinization (ΔHgel) than KP. The swelling power, solubility, amylose content, and transmittance values of KJ and KB potato starches were found to be greater, whereas turbidity values were found to be lower. KP starch showed the highest WAC, while it was lowest for KJ starch, which can be attributed to the variation in granular structure and loose association of amylose and amylopectin molecules. Singh and Kaur [33] reported that large-size fractions from Kufri Sutlej (KS) and Kufri Jyoti (KJ) starch showed the highest retrogradation, while the same fraction from Kufri Chandermukhi (KC) starch showed the lowest retrogradation.

The water absorption capacity (WAC) has been reported to be different for the starches extracted from different varieties. Singh and Kaur [33] reported that Kufri Chandermukhi small granule fraction showed the highest WAC as compared to Kufri Sindhuri and Kufri Jyoti. Kaur *et al.* [29] observed that pasting temperature (PT) of different potato starches ranged from 64.5 to 69.5°C, the highest for Kufri Sindhuri and the lowest for Kufri Bahar. Neeraj *et al.* [11] maximum WAC in Kufri Chipsona-4 followed by Kufri Sindhuri, while it was minimum in Kufri Badshah. Differences in WAC across cultivars may be related to their starch's amylose and amylopectin levels. Additionally, it can be ascribed to granular structural variation. The presence of a significant number of phosphate groups on the amylopectin molecule may help explain the variations in WAC between starches isolated from various potato types [34].

The phosphorous content of extracted starch was found to be different for different potato varieties. It was minimum in Kufri Pushkar followed by Kufri Sindhuri, while it was maximum in Kufri Chipsona-4 [11, 17]. Pineda-Gomez *et al.* [35] studied the starch of potato cultivars growing in the Andean region in the south of Colombia. It was observed that the apparent amylose and phosphorus concentrations of starches extracted from Mambera, Ratona, and Pastusa cultivars of potato were much greater than those recovered from Unica and Roja-huila. Kaur *et al* [29] reported that the phosphorus content of starch granules is positively linked with the phospholipid concentration. Phospholipids are often abundant in phosphorus-rich starch granules, either adhering to their surface or enclosed within.

There were marked differences in swelling power (SP) of extracted starch from different potato varieties. It was maximum in Kufri Chipsona-4 followed by Kufri Bahar, while it was minimum in Kufri Badshah [11, 17]. Additionally, it was observed that Kufri Chipsona-4 included the least amylose and the most phosphorus, indicating a greater swelling power; however, Kufri Badshah contained the most amylose and the least phosphorus, indicating a lesser swelling power. The difference in SP between starches from various potato cultivars indicates that the affective bonding forces between granules are stronger in certain. The swelling ability of starch is directly proportional to its amylopectin concentration, since amylose functions as a diluent and swelling inhibitor.

Starches' swelling ability and solubility are strongly linked. Neeraj *et al.* [11] reported that maximum solubility was for the starch extracted from Kufri Chipsona-4 followed by Kufri Pushkar, while it was minimum in Kufri Badshah. The variations in swelling power and solubility of the starches among different varieties can be attributed to the differences in granule size, amylose content, molecular structure of amylopectin and the crystallinity as well as the granule organization.

Neeraj *et al.* [11] observed maximum light transmittance in starch extracted from Kufri Chipsona-4 followed by Kufri Bahar, while it was minimum in Kufri Pushkar. Higher transmittance led to greater starch paste clarity. The clarity of the paste varied significantly depending on the starch source, amylose/amylopectin ratio, chemical or enzymatic changes, and solute inclusion. Potato cultivars with a substantial increase in transmittance may be used in the jelly, beverage, and fruit paste sectors to get the appropriate consistency.

Syneresis, or freeze-thaw stability, is a critical characteristic used to assess a starch's capacity to survive the undesired physical changes that occur during freezing and thawing. Syneresis was discovered to be least in starch extracted from Kufri Pushkar, followed by Kufri Bahar, and to be most in Kufri Chipsona-4 [11, 17].

Kaur *et al.* [32] reported that endothermic peaks for several potato cultivars ranged from 59.96 to 68.89°C, the peak temperature was 63.37–64.58°C, and the final temperature was 67.4–68.9°C.

4. Effect of curing treatment of tubers

Freshly harvested potatoes have very short shelf life due to thin skin. Curing is accomplished by holding potatoes in dark at ~22°C and RH 90% for 10—15 days. During curing potatoes utilize the reserved food material to provide energy and metabolites to heal bruises and cracks and to develop periderm layer making the peel thick and impermeable to water. The various changes taking place in the characteristics of extracted starch from different varieties of potato due to curing of tubers was studied in detail by Neeraj *et al* [11]. It was reported that curing of tubers resulted in decreased starch yield, amylose content, swelling power, and solubility; but increased phosphorus content, WAC, and light transmittance of the extracted starch. The lower starch content in cured potatoes compared to fresh potatoes may be due to its utilisation in the process of periderm layer formation [36]. Van Der Maarel *et al.* [37] reported a decrease in amylose content in cured potatoes as a result of increased activity of the debranching enzyme glucoamylase, which breaks the -1,6 glycosidic bonds present in amylose to form linear amylose. The amylase, pullulanase, glucoamylase, and isoamylase further breakdown amylose into sugars. The increased phosphorus concentration in cured tubers might be attributed to the poor starch recovery from cured tubers that resulted in a greater phosphorus content per unit of recovered starch. The greater WAC content of cured potato starch can be attributed to its reduced amylose and higher phosphorus content. Liu *et al.* [38] found a negative connection between WAC and amylose and a positive correlation between WAC and phosphorus in starches. Reduced swelling power, solubility, and light transmittance of cured potato starch were attributed to its decreased amylose and high phosphorus concentrations.

Curing did not significantly affect the size of the starch particles and syneresis [39]. The syneresis exhibited a significant positive correlation with amylose content. Though the amylose content was slightly reduced by curing, still the syneresis was not affected [11, 17]. This might be because the amount of the change in amylose concentration was too little to have a meaningful influence on syneresis, or because other variables such as crude fibre, fat, protein, and granular structure also playing a role in syneresis [18].

5. Effect of sprout inhibiting treatments

The sprout inhibiting treatments besides affecting the physiology of tubers also alter the properties of various biochemical constituents. CIPC, also referred as chlorpropham is the most commonly used sprout suppressant on potatoes when stored at 8–12°C. Potatoes can also be stored for at least 12 weeks at either 8 or 18°C without sprouting, if tubers are dipped in hot water (57.5°C and 20 or 30 min) [5, 40]. It was reported by Hu *et al.* [40] that there were no significant differences in the pasting properties and onset (T_O), peak (T_P) and endset (T_E) temperatures of gelatinisation of sweet potato starch among heat treated (HWT) and non heat-treated samples. Peak viscosity decreased gradually and fluctuated around 310–357 RVU in variety Kanoya control samples, while it increased gradually and fluctuated around 209–308 RVU in Kanoya hot water treated samples, indicating that heat treatment reduced the peak viscosity of potato starch.

Potatoes treated with CIPC to inhibit sprouting contained greater amounts of total starch as well as resistant starch (RS) than untreated potatoes tubers [41]. Lu *et al.* [4] investigated potato varieties treated with CIPC and stored at 8°C for 5 months and found that the least shifting of gelatinization peaks to lower temperature for CIPC treated samples. Ezekiel and Singh [42] studied starch properties of potato cultivars treated with CIPC. Swelling volume of starch decreased significantly with CIPC in all varieties.

Neeraj [43] studied the effect of various sprout inhibiting treatments viz., hot water dip treatment (HWT, 57.5 ± 0.1°C for 20 min) and single spray of 50% formulation of CIPC treatments on the characteristics of starch extracted from Kufri Chipsona 4 variety of potato stored at low temperature (12 ± 1°C). It was observed that significantly higher starch yield was observed in CIPC treated tubers followed by hot water treatment and untreated. The CIPC treated tubers retained higher starch than other treatments until the end of the storage period, which can be attributed to inhibited sprouting and low respiration rates. Hot water treatment may also have resulted in higher starch yield because of its inhibitory role on sprouting of tubers. Kyriacou *et al.* [44] also observed that CIPC and HWT retained higher starch concentrations during the storage period.

There was no significant effect of HWT on particle size, however, CIPC treatment significantly increased the percentage of small size particles of starch from LT stored tubers [43]. Ezekiel *et al.* [45] reported that CIPC treated potatoes subsequently stored at 12°C for 90 days showed an increase in the proportion of small size granules. The granule diameter ranged from 18 to 25 μm.

The effects of CIPC and HWT were found to be non-significant with respect to untreated tubers for extracted starch's moisture content, protein content, fat content, crude fiber, ash content, purity, WAC and colour whiteness values; while starch yield, amylopectin, phosphorus content, swelling power, solubility, light transmittance and peak viscosity were higher for starch extracted from tubers treated with CIPC than HWT [43]. The nonsignificant effect on purity of starch was attributed to nonsignificant changes in moisture, fat, crude fibre, protein and ash contents of the starch extracted from tubers subjected to various sprout inhibiting treatments.

Hot water dip treated tubers resulted in significantly lower swelling power than untreated tubers at RT storage, however, significantly higher swelling power was observed in CIPC treated tubers stored at LT [43]. Additionally, Lu *et al.* [4] discovered that tubers treated with CIPC exhibited a greater swelling capacity of the extracted starch. The considerably increased phosphorus content of the CIPC-treated tuber starch might account for its relatively strong swelling power. The lack of internal structure produced by negatively charged phosphate ester groups inside starch granules has been attributed to potato starches' greater swelling potential [46]. Eerlingen *et al.* [47] attributed increased swelling power to the amorphous AM being transformed into a helical shape, increased contacts between AM chains, and altered interactions between crystallites and the amorphous matrix. The development of cross links in the amorphous area and consequent rise in crystallinity during HWT might possibly account for the decrease in swelling power and solubility of the extracted starch [48].

Significantly lower syneresis was observed in the starch extracted from CIPC treated tubers followed by HWT tubers and untreated tubers [43]. Singh *et al.* [14] reported that lower amylose content and high percentage of small size granules caused lower syneresis. Thus, lower amylose content and higher percentage of small size granules may be responsible for lower syneresis exhibited by the starch extracted form CIPC treated potatoes. Hu *et al.* [49] reported that HWT reduced the setback viscosity value of sweet potato starch during storage that decreased its syneresis.

Verma *et al.* [50] observed that increasing gamma (γ) irradiation dosage resulted in a substantial drop in the apparent amylose content, swelling index, enthalpy of gelatinization, transition temperature, and overall crystallinity of the starches. Similarly, increasing the ionising dosage resulted in a substantial decrease in the pasting characteristics (peak, trough, setback, ultimate viscosity, and pasting temperature) of the extracted starches. On the contrary, the enhanced solubility index of the starch was caused by gamma irradiation. Irradiation treatment increased the total free glucose content of potato and lowered the starch thermal transition and pasting temperatures. Starch crystallinity reduced substantially in irradiation potatoes, which may account for the lower resistant starch concentration.

6. Effect of storage conditions and packaging

The starch content of potatoes has been reported to decrease during storage due to conversion of starch to sugar and its utilization in respiration [51]. The rate of starch depletion and sugar buildup vary with cultivar and storage temperature, presumably due to variations in enzyme activities [52]. Johnston *et al.* [53] observed an increase in amylose to amylopectin ratio in the starch granules during storage of tubers. Golachowski [54] studied the properties of starch separated from potato tubers stored at 20°C, 8°C, 4°C and 0°C. The study included also potato tubers having been frozen, thawed and refrozen as well as potato tubers before storage. Starch separated from the stored potato tubers showed differences in chemical composition, reducing power, granularity, whiteness and viscosity as well as gelatinization temperatures in comparison with the starch separated from potato tubers before storage. Ridley and Hogan [55] reported a reduction in the viscosity of starch isolated from potatoes that were stored at 1.7 and 7.2°C for three months. Peak viscosity was reduced after 90 days of storage at 8°C but increased at 16°C, with no significant change at 4 or 12°C. Separated starches from potatoes kept at a higher temperature demonstrated increased transition temperatures and decreased pasting temperatures, and vice versa. Separated starch from potatoes kept at 8°C exhibited increased peak, trough, and breakdown viscosity, as well as a reduced setback. After storage at 4, 8, and 20°C, the fraction of large size granules decreased and the proportion of smaller granules increased.

Cottrell *et al.* [56] investigated the physicochemical properties of starch produced from several potato cultivars and kept at temperatures of 4, 8, 10, and 20°C. At harvest, the surface of the starch granules was smooth but during storage the surface of the granules became progressívely more pitted. Potato starches kept at 4°C had reduced To, Tc, and Hgel values, as well as lower transmittance values. The peak viscosity and setback of starches isolated from potatoes stored at a higher temperature (20°C) were the greatest.

Kaur *et al.* [57] reported that the starches extracted from potatoes kept at 4°C had significantly greater proportion of tiny granules, pitted with rougher surfaces lower commencement and conclusion gelatinization temperatures than those held at 8 and 20°C.

Ezekiel *et al.* [52] detected a substantial reduction in the amylose content of starch with increase storage period of tubers. Separated starch from potatoes that had been kept at a higher temperature demonstrated a lower pasting temperature. Potato starch kept at 8°C had a larger peak, trough, and breakdown viscosity, as well as a lower set-back. Peak viscosity rose as storage temperature increased, whereas swelling volume dropped. After 90 days of storage of tubers at 4°C, a substantial decrease in swelling volume but no significant changes in phosphorus content of the extracted starch were observed.

Neeraj [43] reported that as the storage time for tubers increased, the moisture, fat, ash, crude fibre, amylopectin, phosphorus, water absorption capacity, light transmittance, and peak viscosity of the extracted starch increased; while yield, purity, amylose, swelling power, solubility, syneresis, and colour whiteness value decreasedIt has been reported that pastes prepared from potato starches with higher percentages of small granules exhibit lower syneresis. The decrease in syneresis of starch extracted from stored tubers could be due to the decreased amylose, and increased amylopectin, phosphorus and percentage of small size particles [17, 18].

Neeraj [43] reported that the starch extracted from the tubers stored at low temperature (LT) with different packaging viz., nylon mesh bags or MAP or vacuum did not show significant differences in its moisture, fat, protein, ash, crude fiber, purity, amylose, amylopectin, WAC, swelling power and whiteness values; while minimum syneresis and maximum starch yield, phosphorus content, solubility, light transmittance and peak viscosity were observed for starch extracted from tubers packed in net bags followed by vacuum and modified atmosphere packaging. Significantly higher peak viscosity of starch was observed for starch extracted from net bag packed tubers followed by vacuum, while it was minimum in modified atmospheric packaging. The various packaging methods did not significantly affect the starch yield from tubers stored at room temperature, however, for tubers stored at low temperature, maximum starch yield was observed for tubers packed in net bags, while modified and vacuum packed tubers were showing at par but lower starch yields. Similarly, it was reported by Mare and Modi [58] that Taro cormels of Dumbe-dumbe and Pitshi packaged in mesh bags also displayed higher starch content compared with those packaged in polyethylene bags and boxes.

7. Conclusion

The growing starch markets have led to food industries to a constant demand for starches with specific properties that meet the demands of applicability. The quality of extracted starch form potato tubers, however, are significantly affected by environmental, cultural and storage conditions. The available information of the various physico-chemical and functional characteristics of potato starch as affected by extraction methods, varieties, curing, sprout inhibitors, and storage conditions. Some more information still need to be generated with respect to the effect of preharvest cultural practices, organic raising, extent of sprouting of tubers, application of other sprout suppressants (irradiation, isopropyl phenylcarbamate, ozone, ethylene, MH, carvones, etc) and various storage structures on the quality of extracted potato starch. The information w.r.t. verities may be helpful in identifying phenotyping trait(s) in potato breeding processes to create special potato genotypes for manufacturing of starch with the characteristics specially tuned for certain industrial processing technologies. On the basis of information, the potato starch industry may also select specifically pre- and postharvest treated stored potatoes of a given variety for extracting starches having the desired functional characteristic.

DOI: http://dx.doi.org/10.5772/intechopen.101831

Acronyms and abbreviations

AM	amylose
AP	amylopectin
CIPC	isopropyl N-(3-chlorophenyl) carbamate
HWT	hot water treatment
LT	low temperature
RS	resistant starch
XRD	X-ray diffraction
SP	swelling power
T_E	gelatinization endset temperature
T_O	gelatinization onset temperature
T_P	gelatinization peak temperature
Tvdi	temperature at which viscosity development is initiated
WAC	water absorption capacity
PHI	peak height indices
ΔHgel	enthalpies of gelatinization.

Author details

Saleem Siddiqui[1*], Naseer Ahmed[2] and Neeraj Phogat[3]

1 School of Agricultural Sciences, Sharda University, Greater Noida, Uttar Pradesh, India

2 Department of Food Technology, DKSG Akal College of Agriculture, Eternal University, Baru Sahib, Himachal Pradesh, India

3 Centre of Food Science and Technology, CCS Haryana Agricultural University, Hisar, Haryana, India

*Address all correspondence to: saleem.siddiqui@sharda.ac.in

IntechOpen

© 2022 The Author(s). Licensee IntechOpen. This chapter is distributed under the terms of the Creative Commons Attribution License (http://creativecommons.org/licenses/by/3.0), which permits unrestricted use, distribution, and reproduction in any medium, provided the original work is properly cited.

References

[1] FAO. FAOSTAT Statistical Database. Rome: FAO; 2019

[2] Huang C-C, Lin M-C, Wang C-C. Changes in morphological, thermal and pasting properties of yam (*Dioscorea alata*) starch during growth. Carbohydrate Polymers. 2006;**64**(4): 524-531

[3] Paul V, Ezekiel R, Pandey R. Sprout suppression on potato: Need to look beyond CIPC for more effective and safer alternatives. Journal of Food Science and Technology. 2016;**53**(1):1-18

[4] Lu Z-H, Donner E, Yada RY, Liu Q. Impact of γ-irradiation, CIPC treatment, and storage conditions on physicochemical and nutritional properties of potato starches. Food Chemistry. 2012;**133**(4):1188-1195

[5] Ranganna B, Raghavan G, Kushalappa A. Hot water dipping to enhance storability of potatoes. Postharvest Biology and Technology. 1998;**13**(3):215-223

[6] Boivin M, Bourdeau N, Barnabé S, Desgagné-Penix I. Sprout suppressive molecules effective on potato (*Solanum tuberosum*) tubers during storage: A review. American Journal of Potato Research. 2020;**97**:1-13

[7] Frazier MJ, Olsen N, Kleinkopf G. Organic and Alternative Methods for Potato Sprout Control in Storage. University of Idaho Extension, Idaho Agricultural Experiment Station; 2004

[8] Aina AJ, Falade KO, Akingbala JO, Titus P. Physicochemical properties of Caribbean sweet potato (*Ipomoea batatas* (L) Lam) starches. Food and Bioprocess Technology. 2012;**5**(2):576-583

[9] Ezekiel R, Singh B, Kumar D, Mehta A. Processing quality of potato varieties grown at two locations and stored at 4, 10 and 12°C. Potato Journal. 2007;**34**(3-4):164-173

[10] Villarreal ME, Ribotta PD, Iturriaga LB. Comparing methods for extracting amaranthus starch and the properties of the isolated starches. LWT-Food Science and Technology. 2013;**51**(2):441-447

[11] Neeraj P, Siddiqui S, Dalal N, Srivastva A, Bindu B. Effects of varieties, curing of tubers and extraction methods on functional characteristics of potato starch. Journal of Food Measurement and Characterization. 2021;**14**(6):3434-3444

[12] Lin J-H, C-y L, Chang Y-H. Change of granular and molecular structures of waxy maize and potato starches after treated in alcohols with or without hydrochloric acid. Carbohydrate Polymers. 2005;**59**(4):507-515

[13] Betancur AD, Chel GL, Cañizares HE. Acetylation and characterization of *Canavalia ensiformis* starch. Journal of Agricultural and Food Chemistry. 1997;**45**(2):378-382

[14] Singh J, Singh N. Studies on the morphological and rheological properties of granular cold water soluble corn and potato starches. Food Hydrocolloids. 2003;**17**(1):63-72

[15] Xu X, Huang X-F, Visser RG, Trindade LM. Engineering potato starch with a higher phosphate content. PLoS One. 2017;**12**(1):e0169610

[16] Adebowale KO, Olu-Owolabi BI, Olawumi EK, Lawal OS. Functional properties of native, physically and

chemically modified breadfruit (*Artocarpus artilis*) starch. Industrial Crops and Products. 2005;**21**(3): 343-351

[17] Neeraj P, Siddiqui S, Dalal N, Srivastva A, Pathera AK. Physicochemical, morphological, functional, and pasting properties of potato starch as a function of extraction methods. Journal of Food Measurement and Characterization. 2021;**15**(3): 2805-2820

[18] Raj N, Dalal N, Bisht V, Dhakar U. Potato starch: Novel ingredient for food industry. International Journal of Current Microbiology and Applied Sciences. 2020;**9**(1):1718-1724

[19] Karim A, Toon L, Lee V, Ong W, Fazilah A, Noda T. Effects of phosphorus contents on the gelatinization and retrogradation of potato starch. Journal of Food Science. 2007;**72**(2):C132-C1C8

[20] Chan H-T, Bhat R, Karim AA. Effects of sodium dodecyl sulphate and sonication treatment on physicochemical properties of starch. Food Chemistry. 2010;**120**(3):703-709

[21] Wang S, Li C, Copeland L, Niu Q, Wang S. Starch retrogradation: A comprehensive review. Comprehensive Reviews in Food Science and Food Safety. 2015;**14**(5):568-585

[22] Palacios-Fonseca A, Castro-Rosas J, Gómez-Aldapa C, Tovar-Benítez T, Millán-Malo B, Del Real A, et al. Effect of the alkaline and acid treatments on the physicochemical properties of corn starch. CyTA-Journal of Food. 2013;**11**(Suppl. 1):67-74

[23] Wang L, Wang Y-J. Rice starch isolation by neutral protease and high-intensity ultrasound. Journal of Cereal Science. 2004;**39**(2):291-296

[24] Sajeev M, Moorthy S. Swelling and viscometric characteristics of cassava starch settled in the presence of different chemicals. Trends in Carbohydrate Chemistry. 2000;**7**:117-125

[25] Perera C, Hoover R. Influence of hydroxypropylation on retrogradation properties of native, defatted and heat-moisture treated potato starches. Food Chemistry. 1999;**64**(3):361-375

[26] Lawal OS. Composition, physicochemical properties and retrogradation characteristics of native, oxidised, acetylated and acid-thinned new cocoyam (*Xanthosoma sagittifolium*) starch. Food Chemistry. 2004;**87**(2): 205-218

[27] Khlestkin V, Eltsov I. Different reactivity of raw starch from diverse potato genotypes. Molecules. 2021; **26**(1):226

[28] Mareček J, Frančáková H, Bojňanská T, Fikselová M, Mendelová A, Ivanišová E. Carbohydrates in varieties of stored potatoes and influence of storage on quality of fried products. Journal of Microbiology, Biotechnology and Food Sciences. 2021;**2021**:1744-1753

[29] Kaur A, Singh N, Ezekiel R, Guraya HS. Physicochemical, thermal and pasting properties of starches separated from different potato cultivars grown at different locations. Food Chemistry. 2007;**101**(2):643-651

[30] Singh N, Kaur L, Sandhu KS, Kaur J, Nishinari K. Relationships between physicochemical, morphological, thermal, rheological properties of rice starches. Food Hydrocolloids. 2006; **20**(4):532-542

[31] Singh J, Singh N. Studies on the morphological, thermal and rheological properties of starch separated from some

Indian potato cultivars. Food Chemistry. 2001;**75**(1):67-77

[32] Kaur L, Singh N, Sodhi NS. Some properties of potatoes and their starches II. Morphological, thermal and rheological properties of starches. Food Chemistry. 2002;**79**(2):183-192

[33] Singh N, Kaur L. Morphological, thermal, rheological and retrogradation properties of potato starch fractions varying in granule size. Journal of the Science of Food and Agriculture. 2004;**84**(10):1241-1252

[34] Hoover R. Composition, molecular structure, and physicochemical properties of tuber and root starches: A review. Carbohydrate Polymers. 2001;**45**(3):253-267

[35] Pineda-Gomez P, González NM, Contreras-Jimenez B, Rodriguez-Garcia ME. Physicochemical characterisation of starches from six potato cultivars native to the Colombian andean region. Potato Research. 2021;**64**(1):21-39

[36] Smith O. Effect of cultural and environmental conditions on potatoes for processing. In: Talburt WF, Smith O, editors. Technology of Potato Processing. India: Medtech - A Division of Scientific International; 2018. p. 70-109

[37] Van Der Maarel MJ, Van der Veen B, Uitdehaag JC, Leemhuis H, Dijkhuizen L. Properties and applications of starch-converting enzymes of the α-amylase family. Journal of Biotechnology. 2002;**94**(2):137-155

[38] Liu Q, Weber E, Currie V, Yada R. Physicochemical properties of starches during potato growth. Carbohydrate Polymers. 2003;**51**(2):213-221

[39] Phogat N, Siddiqui S, Dalal N, Srivastva A, Bindu B. Effects of varieties, curing of tubers and extraction methods on functional characteristics of potato starch. Journal of Food Measurement and Characterization. 2020;**14**(6):3434-3444

[40] Hu W, Tanaka S-I, Hori Y. Effect of heat treatment on quality of sweet potato in wrapper type cold store during long-term storage. Japan: Journal of the Faculty of Agriculture-Kyushu University. 2004;**49**(1):129-138

[41] Vaughn KC, Lehnen LP. Mitotic disrupter herbicides. Weed Science. 1991;**39**(3):450-457

[42] Ezekiel R, Singh B. Effect of cooking and processing on CIPC residue concentrations in potatoes and processed potato products. Potato Research. 2007;**50**(2):175-184

[43] Neeraj P. Studies on starch isolated from potato subjected to various storage conditions [Ph.D. thesis (Food Science and Technology)], Hisar, Haryana, India: CCS Haryana Agricultural University; 2017

[44] Kyriacou MC, Gerasopoulos D, Siomos AS, Ioannides IM. Impact of hot water treatment on sprouting, membrane permeability, sugar content and chip colour of reconditioned potato tubers following long-term cold storage. Journal of the Science of Food and Agriculture. 2008;**88**(15):2682-2687

[45] Ezekiel R, Singh B. Changes in contents of sugar, free amino acids and phenols in four varieties of potato tubers stored at five temperatures for 180 days. India: Journal of Food Science and Technology-Mysore. 2007;**44**(5):471-477

[46] Kim YS, Wiesenborn DP, Orr PH, Grant LA. Screening potato starch for novel properties using differential scanning calorimetry. Journal of Food Science. 1995;**60**(5):1060-1065

[47] Eerlingen RC, Jacobs H, Block K, Delcour JA. Effects of hydrothermal treatments on the rheological properties of potato starch. Carbohydrate Research. 1997;**297**(4):347-356

[48] Sharma M, Yadav DN, Singh AK, Tomar SK. Rheological and functional properties of heat moisture treated pearl millet starch. Journal of Food Science and Technology. 2015;**52**(10):6502-6510

[49] Hu W, Jiang A, Jin L, Liu C, Tian M, Wang Y. Effect of heat treatment on quality, thermal and pasting properties of sweet potato starch during yearlong storage. Journal of the Science of Food and Agriculture. 2011;**91**(8):1499-1504

[50] Verma K, Jan K, Bashir K. γ irradiation of cowpea and potato starch: Effect on physicochemical functional and rheological properties. Journal of Food Processing & Technology. 2019;**10**:810

[51] Smith AM, Zeeman SC, Smith SM. Starch degradation. Annual Review of Plant Biology. 2005;**56**:73-98

[52] Ezekiel R, Rana G, Singh N, Singh S. Physico-chemical and pasting properties of starch from stored potato tubers. Journal of Food Science and Technology. 2010;**47**(2):195-201

[53] Johnston F, Urbas B, Khanzada G. Effect of storage on the size distribution and amylose/amylopectin ratio in potato starch granules. American Potato Journal. 1968;**45**(9):315-321

[54] Golachowski A. Properties of starch obtained from potato tubers influenced by various temperatures. Starch-Stärke. 1985;**37**(8):263-266

[55] Ridley S, Hogan J. Effect of storage temperature on tuber composition, extrusion force, and Brabender viscosity. American Potato Journal. 1976;**53**(10):343-353

[56] Cottrell J, Duffus C, Mackay G, Allison M. Changes in the surface morphology of starch granules of the cultivated potato, *Solanum tuberosum* L. during storage. Potato Research. 1993;**36**(2):119-125

[57] Kaur A, Singh N, Ezekiel R, Sodhi NS. Properties of starches separated from potatoes stored under different conditions. Food Chemistry. 2009;**114**(4):1396-1404

[58] Mare R, Modi A. Taro corm quality in response to planting date and post-harvest storage: I. Starch content and reducing sugars. African Journal of Agricultural Research. 2012;**7**(19):3014-3021

Chapter 6

Comparative Study of the Physiochemical Composition and Techno-Functional Properties of Two Extracted Acorn Starches

Youkabed Zarroug, Mouna Boulares, Dorra Sfayhi and Bechir Slimi

Abstract

Due to the increase of search for new promising ingredients with interesting properties to develop new industrial food products, the valorization of undervalued resources became a challenge. Considering this, various species of genus *Quercus* acorns represent new resources of highly-valued food ingredients such as starch which encourage its extraction and valorization in food industries. In this regard, collected data from the literature provide an evidence review on the physiochemical and techno-functional properties of different acorn starches extracted from Tunisian species, especially; *Quercus ilex* L. and *Quercus suber* L. The reported data on X-ray diffraction analysis are, also, discussed. Data highlighted the possibility of using the extracted *Quercus* starches to develop new functional food products and improve technological properties and shelf life of products solicited by consumers.

Keywords: acorn starch, physiochemical composition, techno-functional properties, X-ray diffraction analysis

1. Introduction

Genus *Quercus* acorns belong to the family *Fagaceae*, which includes several species such as *Quercus robur, Quercus petraea, Quercus suber, Quercus ilex*, and *Quercus pubescens* [1]. These species produce a widely known fruit, named acorns, which are of vital importance for both humans and animals. The acorn fruits composition varies with species and origin. Acorn fruits are good source of carbohydrates (starch), proteins, fats, minerals, essential amino acids, vitamins (mostly A and C), unsaturated fatty acids (oleic acid) and sterols [2]. The nutritional composition of acorn fruit is comparable to many cereal grains. Moreover, acorns are a good source of active compounds, such as phenolic acids, and flavonoids, with an interesting antioxidant activity [1]. Acorns also contain a high content of tannic acid, a mild toxin giving them a bitter taste that can be removed by many methods such as soaking in water, boiling, or roasting [3]. Traditionally, acorns were used in the human diet, generally as flour for bread production, or as a coffee substitute beverage (after a roasting process) [4]. Recently, acorn fruit flour was included in many other food preparations such as pasta, biscuits, hot beverages, cakes, and cookies [5]. Among all the nutrients present in acorn fruit, starch was the predominant component with content ranging from 31 to 55% [1]. Thus, further interest must be given to acorn starch extraction and valorization. Starch is a biodegradable carbohydrate polymer which has been widely studied due to its availability, price, and extensive industrial use. Recent research has shown that acorn starch can be used as an ingredient for food and nonfood applications [6, 7]. Acorn starch was used in many industrial applications like in the cases of paper, plastics, textile, pharmaceutical, and cosmetic industries [8]. Also, starch is a raw material representing the principal component of many food formulations being responsible for important functional and textural properties and nutritional characteristics of the many food products [9]. Owing to its interesting properties such as high resistance and paste consistency, acorn starch can be used as thickening and stabilizing agents in food formulations [2, 7]. The chemical composition and physicochemical features of starch are mainly characteristic of their biological origin. Starch from all plant sources has many similar properties but they do also differ in many aspects. Variation in structure, crystallinity, chemical composition and functional properties of starch granules are depending on their botanical origin and growth conditions. For the selection of the specific use of acorn starch, it is necessary to understand the physicochemical and functional properties of extracted starches from various acorn species. From all the above, collected data related to the chemical and technological properties of acorn starches extracted from *Q. ilex* L. and *Q. suber* L. species as well as characteristics of other starches obtained from various botanical sources such as corn, potato and cassava are discussed to highlight the importance to valorize acorn starches and their potential applications. Thus, it will be interesting to valorize acorn known as a Tunisian under valuated resource by providing promising new ingredients to formulate new food products.

2. Starch extraction methods

Generally, fresh acorn fruits were manually collected from the North West of Tunisia. *Q. ilex* L. is abundant in Bizerte (north east of Tunisia) region, while, *Q. suber* L. is provided from the mountainous region of Ain Drahem from Jendouba in the north west of Tunisia. Due to their short shelf life, acron fruits are, first, hand-peeled, dried in mild conditions and then milled into fine flour.

The extraction technologies of acorn starch consist of dry and wet methods. The use of dry methods is shown unsucceful for the elimination of protein, fat, and tannins from acorn flour, which need to use some other absorbents.

The acorn flour is used for starch extraction using different methods as alkaline washing, hot-water soaking, ultrasonic-assisted ethanol soaking and ultrasonic-assisted hot-water soaking [10]. The three later methods lead to starch granules with similar internal structure. However, starch granules isolated using hot-water methods are complete and glossy with a few numbers of pits. It's important to know that the ultrasonic technology became the most effective in food applications compared with conventional technologies.

3. Starch extraction yield

Numerous studies have already been conducted on the starch yield of acorn species originating from countries all over the world. It is stated in previous studies that starch is the main component of acorns and usually constitute more than 50% of the kernel [11]. The yield of starch extracted from acorn species cultivated in Tunisia and other countries of the world as reported in different studies is presented in **Table 1**. The starch yield varied from 17.3 to 89.83% in acorn species. The starch content in *Q. ilex* L. and *Q. suber* L. were reported to vary from 48.93 to 89.83% [14] and 86.9 to 88.5% [12], respectively. The values indicated that acorn fruits are promising crops as an alternative source of starch. In general, the authors concluded that starch content in *Q. suber*. L is higher than that in *Q. ilex* L. reporting a starch yield of about 34.5% [7]. As it can be also seen from **Table 1** that *Quercus palustris Muenchh* [15] contain the lowest amount of starch (17.3%) as compared to other acorn species. Correia et al. [13] and Masmoudi et al. [17] reported values of 88.5% and 63% of starch content in the Portuguese and Tunisian *Q. suber* L. fruits.

Irinislimane and Belhaneche Ben semr [18] and Correia et al. [12] isolated starch from *Quercus Suber* L. acorns and observed a starch yield accounting 21% and 31.4%, respectively. The starch content in potatoes, tubers and roots are reported to vary from 75 to 80% and 75.4 to 77.4% [5], respectively.

This variability in the starch yield was due to the difference in plant species, cultivation climate, ripening stage, harvesting time of fruits, and extraction method used [7].

The obvious retained conclusion is that the high content of starch makes the *Quercus* species particularly *Q. suber* L. ideal for starch extraction and valorization in many food and nonfood industries applications. Besides, the content of starch in acorn flour gave it good functional characteristics related to starch such as viscosity, swelling, and gelling [17]. In fact, it is suggested that acorn starch might be used as thickening and stabilizing agent, owing to its high paste consistency [2]. Since this polysaccharide is present as resistant starch in a high percentage, it can be very useful as a prebiotic growth promoter, constituting a good alternative to other current prebiotic agents such as fructo-oligosaccharides, inulin, isomalto-oligosaccharides, polydextrose, and lactulose [19].

Several studies have examined the effect of different methods using both physical and chemical methods on acorn starch yield. Differences in starch content are

Species	Starch yield (%)	Extraction methods	References
Quercus suber	86.9	Enzymatic treatment	[12]
	88.5	Alkaline method	[12]
	49	Alkaline method	[13]
	45–57	Alkaline method	[12]
Quercus ilex	34.5	Alkaline method	[7]
	48.93–89.83	Alkaline method	[14]
Quercus palustris Muenchh	17.3	Water method	[15]
Quercus leucotrichophora	54.7	Alkaline method	[16]
Quercus rotundifolia	48	Alkaline method	[13]

Table 1.
Extraction methods and starch yield of various acorn species.

observed using alcohols-based extraction, alkaline-based extraction, acetone-based extraction, hot-water soaking, ultrasonic-assisted ethanol soaking and ultrasonic-assisted hot-water soaking [10].

4. Physico-chemical composition

The physico-chemical composition of acorn starch extracted from *Quercus* species is affected by extraction and purification methods and the origin of raw materials [7]. The extracted acorn starch contained water and minor components such as lipids, proteins and ash (**Table 2**). The moisture content of acorn starch species varies from 7.22 to 15.91%. The moisture content of *Q. ilex* L. extracted starch from four different areas in Algeria varied from 2.2% to 15.9% [14]. The moisture content is very important parameters for the determination of the starch quality. A low moisture content of the acorn starch less than 20% is acceptable for commercial starch and a value less than 13% is recommended for safe storage [9]. Such values are close to those reported in cereal (10–12%) and some roots and tubers (14–18%) starches [20].

Several studies show low lipids, proteins, and ash contents in starches extracted from different acorn species.

Lipids have an essential role in the properties of starch, which is associated with the textural properties of various foods. The lipid content in all starches extracted from *Quercus* species is below 1%. Tunisian *Q. ilex* L. starch contains a high amount of lipid content (0.51%) than both potato and wheat starches [21].

The ash content of the extracted starch from *Q. ilex* grown in Tunisia (20.66%) is relatively high. The review of Taib and Bouyazza, [5], reported ash values ranged from 0.01 to 1.41% in different *Quercus* species. The low ash content illustrates the purity of starch after the extraction and isolation processes. The protein contents of starches obtained from different *Quercus* species ranged from 0.25 to 1.05%. These low values show a high extracted starch purity and quality [7]. The pH values in *Q. suber* was about 5.6 units, while it ranged between 4.73 and 6.43 units in *Q. ilex* starches. Such value is lower than that (6.22 units) reported for the potato starch [22]. High pH value could lead to undesirable protein modification as well as molecular

Components	***Quercus suber*. L**		***Quercus ilex*. L**	
	[7–12, 14]	**[12]**	**[7]**	**[14]**
Moisture (%)	12.95	7.22	10.17	2.2–15.9
Fat (%)	0.50–0.60	Nd	0.51	0.23–0.63
Protein (%)	0.25–0.3	Nd	0.92	0.91–1.05
Ash (%)	0.02–0.24	0.13	2.66	0.14–0.18
pH	5.60	Nd	5.3	4.73–6.43
L*	61.13	Nd	85.03	Nd
a*	0.84	Nd	0.52	Nd
b*	15.07	Nd	10.2	Nd

Nd: not determined.
L, a* and b* are the color parameters.*

Table 2.
Physico-chemical composition of acorn species starches.

cross linkage and rearrangements resulting in the formation of toxic compound [23]. A positive correlation was obtained between the pH value, the fat and the protein contents [24]. The variation observed in the chemical composition of starches is assigned to the extraction and purification methods, environmental conditions (climate and soil composition), growth stage of plant and genotype. During different seasons of the year (summer and autumn), protein and fat contents vary in four collected acorn species (*Q. suber*, *Q. ilex*, *Q. faginea* and *Q. pyrenaica*) [25]. From these findings, we conclude that the chemical content of starches is influenced by the botanical source and the extraction methods used.

Concerning the color parameters, the extracted starches from Tunisian *Quercus* species exhibited a slightly yellow-white color. Indeed, extracted *Q. ilex* L. starch showed a high lightness L* value (85.03) [7] compared to the *Q. suber* L. (61.13). While, the obtained values of a* (0.52) and b* (10.2) were lower than those found in *Q. suber*. L (0.84 and 15.07, respectively) [7]. These findings showed that acorn fruit is a good source of starch that can be used in food industry without the necessity of chemical or genetic modifications. This polysaccharide may be industrially applied as emulsifiers, stabilizers, and thickeners in food as well as prebiotic growth promoter [26].

5. Swelling power, solubility and water absorption

When starch is heated in excessive amount of water, its crystalline structure is disrupted, and water molecules become linked by hydrogen bonding to the exposed hydroxyl groups of amylose and amylopectin [5]. These phenomena results in the swelling, solubility and increasing volume of starch granules. The swelling power, solubility, and water absorption values of extracted acorn starches from *Q. ilex* and *Q. suber* species are presented in **Table 3**. These parameters increased progressively with the increase of temperature from 60 to 90°C. The solubility of the extracted *Q. ilex* starch is higher than that of *Q. suber*, and ranged from 0.2 to 12.95% at 60°C and 4–64.22% at 90°C.

However, the swelling power and the water absorption values are lower in *Q. ilex* starch compared to the *Q. suber* starch. Diversity in swelling, solubility and water absorption of acorn starches has been observed. Boukhelkhal and MoulaiMostefa. [14] reported low solubility and swelling power of four species of acorn starches at temperature of 90°C ranging from 4 to 14% and 11 to 13 g/g, respectively. Values related to the swelling power are low compared to those found by Singh et al. [27] and Elmi Sharlina

	***Quercus ilex*. L starch [7–11, 14]**				***Quercus suber*. L starch**			
Temperature (°C)	**60**	**70**	**80**	**90**	**60**	**70**	**80**	**90**
Solubility (%)	0.2–12.95	1.8–16.35	2.4–29.28	4–64.22	3.3	4.4	7.5	20.2
Swelling power (g water/g starch)	3.9–8.95	8.4–10.53	10–13.3	11–20.76	6	12.5	13.03	21.51
Water absorption (g water/g starch)	3.68	5.7	9.3	10.1	4.5	8.5	11.6	15

Table 3.
Swelling power solubility and water absorption of acorn species starches.

et al. [22] on sweet potato starch (35 to 40 g/g) and chestnut starches (13.6–17.3 g/g), respectively. These low values are attributed to the amylose content of the acorn starch species [28], starch molecule's ability to hold water, hydrogen bonding, and the degree of crystallinity [29]. According to Jiang et al. [30], the solubility values of starches extracted from five different *Dioscorea* L. species, which are *D. opposite Thunb, D. alata Linn, D.nipponica Makino, D. bulbifera Linn* and *D.septemloba Thunb*, varied from 11.14 to 30.04% at the temperature of 95°C. A comparative study showed that swelling power and solubility of acorn starch at 90°C were higher than those of black wheat, buckwheat, coix and naked oat starches and lower than those of corn, jiaoyu, kuzhu, and longya lily starches [31].

The solubility suggests that additional interactions may have occurred between amylose-amylose and amylopectin-amylopectin chains [32]. Concerning, the water absorption capacity of starch, it corresponds to the hydrogen bonding between water molecules and hydroxyl groups in the starch molecules and starch chains as well as diversification of the starch granule structures [33]. In general, the starch extraction methods have important effect on swelling power, water absorption and solubility parameters of starch. Zhang et al., [10] reported a relatively higher value of swelling power (24.99 g/100 g) and solubility (15.22%), at temperature of 90°C, in acorn starch extracted by an ultrasonic-assisted ethanol soaking method. Variation of these parameters in extracted starches is associated to various factors such as: amylose content, granule size, structure of starch granules, viscosity patterns, and presence of non-starch compounds (lipids, ash and proteins) [5].

6. Refrigeration and freezing stability

In order to evaluate the stability of starch during storage, it was necessary to verify the expulsion of water, expressed by syneresis, contained in gels as a consequence of the reorganization of starch molecules [34]. Collected syneresis values during refrigeration and freezing time are grouped in **Table 4**. Results showed that *Q. suber* starch lost less water than the *Q. ilex* starch under refrigeration (4°C) with the increase of the storage time. However, the latter presents higher syneresis values already from the first freeze time, showing low stability to freezing under the conditions used in the studies and a richness in amylose content [9]. It is known from the literature, that starches with high amylose content such as potato (20.1–31.0%), maize (22.4–32.5%), taro (28.7–29.9%), and cassava (18.6–23.6%) present high syneresis, due to the large amount of water expelled during the retrograding process [25]. It known that during freezing of the starch paste, the cohesive portion of the starch formed a layer and the rest separated into a water layer.

Time (h)	Syneresis to refrigeration 4°C (%)		Syneresis to freezing −20°C (%)	
	Quercus ilex.L starch [7]	*Quercus suber.L* starch	*Q. ilex.L* starch [7]	*Q. suber.L* starch
24	42.81	42	26	46.1
48	49.34	43.44	36.18	47.5
72	53.3	45.1	38.47	55.1
96	55.57	46	41.43	60.2

Table 4.
Refrigeration and freezing stability of acorn species starches.

7. X-ray diffraction analysis

Starch is a semi-crystalline material affected by amylose content and amylopectin chain length that consists of amorphous and crystalline regions. The amylose content directly affects the crystallinity degree of the starch, such that when there is a lack of amylose content, the crystallinity degree increases, whereas the longer chain amylopectin forms have a more stable crystalline structure [31].

Generally, starch is present in three different crystalline structures which are A-type, B-type, and C-type that depended essentially on the variety of starch source.

The difference between A- and B-types of starch granules is due to the arrangement of double helices. A-type starches form dense packing with four water molecules, whereas B-type starch is more open causing more water molecules (36 water molecules) to be located in the center of a hexagonal packing of helices. For this reason, it is indicated that the A-type is more stable and requires a higher temperature than B-type starch for gelatinization [31].

X-ray diffraction analysis was employed to observe the changes in the degree of crystallinity of starch as a result of gelatinization. **Figure 1** resume the X-ray diffraction patterns observed on acorn starches extracted from *Q. suber* and *Q. ilex*. The X-ray diffraction patterns provide a classification of the two acorn starches under an A-type crystalline structure, which characterized most cereal starches [35] showing two strong reflections at 15.2° and 22.7°. The X-ray diffraction patterns of both acorn starches showed four intense diffraction peaks at 15.2°, 17.2°, 19.52°, and 22.7° of 2Θ. The strong reflections at 15.2° and 22.7° of 2θ were classified as the A-type crystalline structure, which characterized most cereal starches [35]. The observed peaks at approximately 17.2° and 19.52° of 2θ were characterized as the B-type pattern. However, the C-type crystalline structure consisted of A- and B-type crystallites. Thus, the X-ray diffraction pattern can contain various superpositions of the characteristic diffraction peaks depending on the ratio between the contents of these polymorphs [35]. In general, cereal starches have an A-type pattern, whereas tuber starches display the B-type pattern, and certain roots and legumes starches show a C-type pattern [36]. Such results were close to those found on starch from *Quercus glandulifera* Bl. [37] and *Dioscorea pyrifolia* tubers [2]. Numerous studies have already

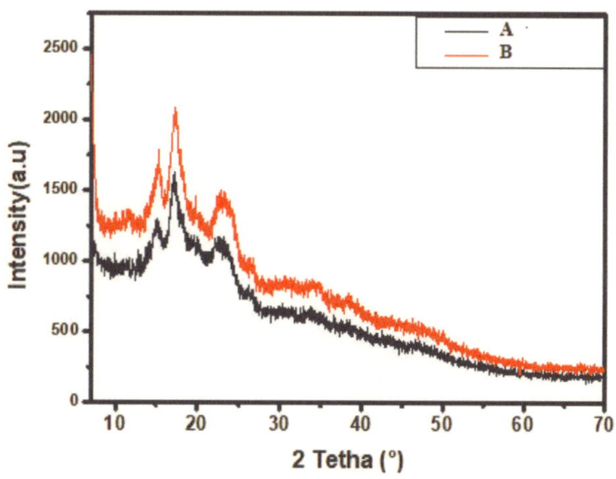

Figure 1.
X-ray diffraction pattern of different starches. A: Quercus ilex. *L starch, B:* Quercus suber. *L starch.*

been conducted on the X-ray diffraction of starches extracted from acorn species originating from countries all over the world. Deng et al. [26] reported that acorn starch from china was B-type. A-type polymorph was reported for acorn starch [38]. However, Zhang et al. [10] and Molavi et al. [39] noted also a C-type polymorph in acorn starches. The difference in the diffraction pattern of starch granules was mainly influenced by genotypic, agronomic, and growing conditions such as temperature [7]. According to Dereje, [40] the type of crystallinity of the extracted starch was influenced by growth temperature, alcohols, fatty acids, and the chain length of amylopectin.

8. Conclusion

Despite that acorns are underutilized fruits, they represent a good alternative source of starch. The acorn starch yield differs from one specie to another representing about 50%. It can be extracted using various methods. The acorn starch was characterized by a yellow color and good technological properties allowing its use during manufacturing of food products. Thus, acorn starch can represent an interesting functional ingredient capable to improve the properties of the final product.

Acknowledgements

This work was supported by the Ministry of Higher Education and Scientific Research Tunisia and the Ministry of Agriculture, Water Resources and Fisheries, Tunisia.

Conflict of interest

The authors declare no conflict of interest.

Author details

Youkabed Zarroug[1*], Mouna Boulares[2], Dorra Sfayhi[1] and Bechir Slimi[3]

1 University of Carthage, Field Crops Laboratory (LR16INRAT02), National Agronomic Research Institute of Tunisia (INRAT), Ariana, Tunisia

2 University of Carthage, Research Unit: Bio-preservation and Valorization of Agricultural Products, Higher Institute of Food Industries (ESIAT), Tunis, Tunisia

3 Laboratoire des Nanomatériaux et Systèmes pour les Energies Renouvelables (LANSER), Centre de Recherches et des Technologies de l'Energie Technopole Borj Cedria, Hammam Lif, Tunisia

*Address all correspondence to: zarrougyoukabed@yahoo.fr

IntechOpen

© 2022 The Author(s). Licensee IntechOpen. This chapter is distributed under the terms of the Creative Commons Attribution License (http://creativecommons.org/licenses/by/3.0), which permits unrestricted use, distribution, and reproduction in any medium, provided the original work is properly cited. (cc) BY

References

[1] Korus J, Witczak M, Ziobro R, Juszczak L. The influence of acorn flour on rheological properties of gluten free dough and physical characteristics of the bread. European Food Research and Technology. 2015;**240**(6):1135-1143

[2] Vinha AF, Barreira JCM, Costa ASG, Oliveira MB, Beatriz PP. A new age for *Quercus* spp. fruits: Review on nutritional and phytochemical composition and related biological activities of acorns. Comprehensive Reviews in Food Science and Food Safety. 2016;**15**(6):947-981

[3] Salkova T, Divisova M, Kadochova S, et al. Acorns as a food resource. An experiment with acorn preparation and taste. Interdisciplinaria Archaeologica Natural Sciences in Archaeology. 2011;**II**(2):133-141

[4] Rakić S, Povrenović D, Tešević V, Simić M, Maletić R. Oak acorn, polyphenols and antioxidant activity in functional food. Journal of Food Engineering. 2006;**74**(3):416-423

[5] Taib M, Bouyazza L. Composition, physicochemical properties, and uses of acorn starch. Journal of Chemistry. 2021;**2021**:9. DOI: 10.1155/2021/9988570

[6] Ozunlu O, Ergezer H, Gokçe R. Improving physico-chemical, antioxidative and sensory quality of raw chicken meat by using acorn extracts. LWT. 2018;**98**:477-484

[7] Zarroug Y, Boulares M, Mejri J, et al. Extraction and characterization of Tunisian *Quercus ilex* starch and its effect on fermented dairy product quality. International Journal of Analytical Chemistry. 2020;**2020**:9. DOI: 10.1155/2020/8868673

[8] Rodrigues A, Emeje M. Recent applications of starch derivates in nanodrug delivery. Carbohydrate Polymers. 2012;**87**:987-994

[9] Pérez-Pachecoa E, Moo-Huchin VM, Estrada-León RJ, Ortiz-Fernández A, MayHernández LH, Ríos-Soberanis CR, et al. Isolation and characterization of starch obtained from *Brosimum alicastrum* Swarts Seeds. Carbohydrate Polymers. 2014;**101**:101920-101927. DOI: 10.1016/j.carbpol.2013.10.012

[10] Zhang Z, Saleh ASM, Wu H, Gou M, Liu Y, Jing L, et al. Effect of starch isolation method on structural and physicochemical properties of acorn kernel starch. Starch–Stärke. 2019; **72**(1-2):1900122

[11] Rababah T, Ereifej K, Al-Mahasneh M, Alhamad M, Alrababah M, Al-u'datt M. The physicochemical composition of acorns for two Mediterranean *Quercus* species. The Journal of Agricultural Science. 2008;**4**:131-137

[12] Correia PR, Nunes MC, Beirão-da-Costa ML. Effect of starch isolation method on physical and functional properties of Portuguese nut starches. II. *Q. rotundifolia* lam. and *Q. suber lam.* acorns starches. Food Hydrocolloids. 2013;**30**(1):448-455

[13] Correia PR, Leitao AE, Beirao-da-Costa ML. Effect of drying temperatures on chemical and morphological properties of acorn flours. International Journal of Food Science & Technology. 2009;**44**:1729-1736

[14] Boukhelkhal M, Moulai-Mostefa N. Physicochemical characterization of starch isolated from soft acorns of holm

oak (*Quercus Ilex Subsp. Ballota* (Desf.) Samp.) grown in Algeria. Journal of Food Measurement and Characterization. 2017;**11**(4):1995-2005

[15] Stevenson G, Jane JL, Inglett GE. Physico-chemical properties of pin oak (*Quercus palustris muenchh.*) acorn starch. Starch-Starke. 2006;**58**(11):553-560

[16] Soni PL, Sharma H, Dun D, Gharia MM. Physicochemical properties of *Quercus leucotrichophora* (Oak) starch. Starch/Stärke. 1993;**45**:127-130

[17] Masmoudi M, Besbes S, Khlifi M, Yahyaoui D, Attia H, Hamadi A. Optimization of acorn (*Quercus suber* L.) muffin formulations: Effect of using hydrocolloids by a mixture design approach. Food Chemistry. 2020;**328**: 127082

[18] Irinislimane H, Belhaneche-Bensemra N. Extraction and characterization of starch from Oak acorn, sorghum, and potato and adsorption application for removal of maxilon red GRL from wastewater. Chemical Engineering Communications. 2017;**204**(8):897-906

[19] Siro I, Kapolna E, Kapolna B, Lugasi A. Functional food. Product development, marketing and consumer acceptance—A review. Appetite. 2008;**51**:456-467

[20] Wani IA, Sogi DS, Hamdani AM, Gani A, Bhat NA, Shah A. Isolation, composition, and physicochemical properties of starch from legumes: A review. Starch-Starke. 2016;**68**(9-10): 834-845

[21] Jiang Q, Liang S, Zeng Y, Lin W, Ding F, Li Z, et al. International Journal of Biological Macromolecules. 2019;**125**:1147

[22] Elmi Sharlina MS, Yaacob A, Lazim A, et al. Physicochemical properties of starch from *Dioscorea pyrifolia* tubers. Food Chemistry. 2017;**220**:225-232

[23] Muhammad U, Tahir I, Raza MS, Muhammad I, Bushra I. Alkaline extraction of starch from broken rice of Pakistan. International Journal of Innovation and Applied Studies. 2014;7(1):146-152

[24] Awoyale W, Sanni LO, Shittu TA, Adebowale AA, Adegunwa MO. Development of an optimized cassava starch-based custard powder. Journal of Culinary Science & Technology. 2017:1-23. DOI: 10.1080/15428052.2017.1404534

[25] Canellas I, Roig S, San MA. In: Robles AB, Ramos ME, Morales MC, Simon E, Gonzalez-Rebollar JL, Boza J, editors. Caracterizacion y evolucion anual del valor bromatologico de las quercıneas mediterraneas. Granada: Pastos, desarollo y conservacion; 2003. pp. 455-462

[26] Deng M, Reddy CK, Xu B. Morphological, physicochemical, and functional properties of underutilized starches in China. International Journal of Biological Macromolecules. 2020; **158**:648-655

[27] Singh V, Ali SZ, Somashekar R, Mukherjee PS. Nature of crystallinity in native and acid modified starches. International Journal of Food Properties. 2006;**9**:845-854

[28] Kaur L, Singh J, Singh N. Effect of cross-linking on some properties of potato (*Solanum tuberosum* L.) starches. Journal of the Science of Food and Agriculture. 2006;**86**(12):1945-1954

[29] Correia PR, Beirão-da-Costa ML. Starch isolation from chestnut and acorn

flours through alkaline and enzymatic methods. Food and Bioproducts Processing. 2012;**90**(2):309-316

[30] Jiang Q, Gao W, Li X, et al. Characterizations of starches isolated from five different *Dioscorea* L. species. Food Hydrocolloids. 2012;**29**(1):35-41

[31] Thanyapanich N, Jimtaisong A, Rawdkuen S. Functional properties of banana starch (*Musa* spp.) and its utilization in cosmetics. Molecules. 2021;**26**:3637. DOI: 10.3390/molecules 26123637

[32] Hughes T, Hoover R, Liu Q, Donner E, Chibbar R, Jaiswal S. Composition, morphology, molecular structure, and physicochemical properties of starches from newly released chickpea (*Cicer arietinum* L.) cultivars grown in Canada. Food Research International, Barking. 2009;**42**(5-6):627-635

[33] Dome K, Podgorbunskikh E, Bychkov A, Lomovsky O. Changes in the crystallinity degree of starch having different types of crystal structure after mechanical pretreatment. Polymers. 2020;**12**(3):641. DOI: 10.3390/polym12030641

[34] Ojogbo E, Ogunsona EO, Mekonnen TH. Chemical and physical modifications of starch for renewable polymeric materials. Materials Today Sustainability. 2020;**7-8**:100028

[35] Ovando-Martınez M, Osorio-Dıaz P, Whitney K, Bello-Perez LA, Simsek S. Effect of the cooking on physicochemical and starch digestibility properties of two varieties of common bean (*Phaseolus vulgaris* L.) grown under different water regimes. Food Chemistry. 2011;**129**(2):358-365

[36] Zhang P, Whistler RL, Be Miller JN, Hamaker BR. Banana starch: Production, physicochemical properties, and digestibility. Carbohydrate Polymers. 2005;**59**:443-458

[37] Singh N, Singh J, Kaur L, Singh Sodhi N, Singh GB. Morphological, thermal and rheological properties of starches from different botanical sources. Food Chemistry. 2003;**81**(2):219-231

[38] Chen LX, Shi X, Li L. Analysis on basic physicochemical properties and antioxidant activities of the starch from acorn. Hans Journal of Food and Nutrition Science. 2019;**8**(3):195-207

[39] Molavi H, Razavi SMA, Farhoosh R. Impact of hydrothermal modifications on the physicochemical, morphology, crystallinity, pasting and thermal properties of acorn starch. Food Chemistry. 2018;**245**:385-393

[40] Dereje B. Composition, morphology and physicochemical properties of starches derived from indigenous Ethiopian tuber crops: A review. International Journal of Biological Macromolecules. 2021;**187**:911-921

Section 3

Recent Advances

Chapter 7

Binary Interactions and Starch Bioavailability: Critical in Limiting Glycemic Response

Veda Krishnan, Monika Awana, Debarati Mondal, Piyush Verma, Archana Singh and Shelly Praveen

Abstract

Limiting starch bioavailability by modifying food matrix dynamics has evolved over the decade, which further envisions low glycemic starch prototypes to tackle chronic hyperglycemia. The dense matrix of whole grain foods like millets and cereals act as a suitable model to understand the dynamics of binary food matrix interactions between starch-lipid, starch-protein & starch-fiber. The state and types of matrix component (lipid/protein/fiber) which interact at various scales alters the starch micro configuration and limits the digestibility, but the mechanism is largely been ignored. Various in-vitro and in-vivo studies have deciphered the varied dimensions of physical interactions through depletion or augmentation studies to correlate towards a natural matrix and its low glycemic nature. The current chapter briefly encompasses the concept of food matrix types and binary interactions in mediating the glycemic amplitude of starch. We comprehensively elaborated and conceptually explained various approaches, which investigated the role of food matrices as complex real food systems or as fundamental approaches to defining the mechanisms. It's a fact that multiple food matrix interaction studies at a time are difficult but it's critical to understand the molecular interaction of matrix components to correlate in-vivo processes, which will assist in designing novel food prototypes in the future.

Keywords: starch, digestibility, food matrix, binary interactions, glycemic response

1. Introduction

Starch is one of the major constituents of reserve food material, which serves as fuel for the human body. The calorific value of starch is 17.5 kJ/g, which is not only responsible for most of the metabolic functions but also acts as a crucial regulatory adjunct to control energy balance. Starch existed as the major dietary nutrient since time immemorial but the dietary transition with enriched refined products as well as carbaholic staples led to the unprecedented increase in the pre-diabetic and diabetic population with characteristic chronic hyperglycemia. Hence glycaemic response (GR) eliciting potential of food known as the glycemic index (GI) or glycemic potential (GP) are major aspects to understand as well as to fine-tune. In a food matrix, starch bioavailability is modulated by the microstructure (cell wall, membrane, cell layers, granular size, etc.) as well as its dense composition (macro and micronutrients) [1]. Based on the interacting components food matrix interactions are classified as binary (two-component), ternary (three-component), and quaternary (four-component) [2–6]. Types of binary interactions and their effect on starch bioavailability are depicted in **Figure 1**.

Among these, a binary component has gained great importance and has been extensively characterized by component depletion or addition studies under in-vitro conditions [7, 8]. The observed low GP of whole grain foods like millets, pigmented rice has been well correlated with such matrix interactions present endogenously, while high GI has also been reported to lower by exogenous addition of such matrix components [9–11]. The state and types of matrix component (lipid/protein/fiber) which interacts at various scales have also been known to alter the starch micro configuration (repeat, reconstruct the sentence limiting the digestibility, result in lowering the glycemic response [9, 12].

Binary interactions have been majorly characterized using nutrient-sensing fluorescent probe-based confocal laser scanning microscopy (CLSM), where the proximity as well as encapsulating effect of matrix components limiting the starch hydrolytic metabolic enzymes have been observed [13, 14]. Further, the effect of

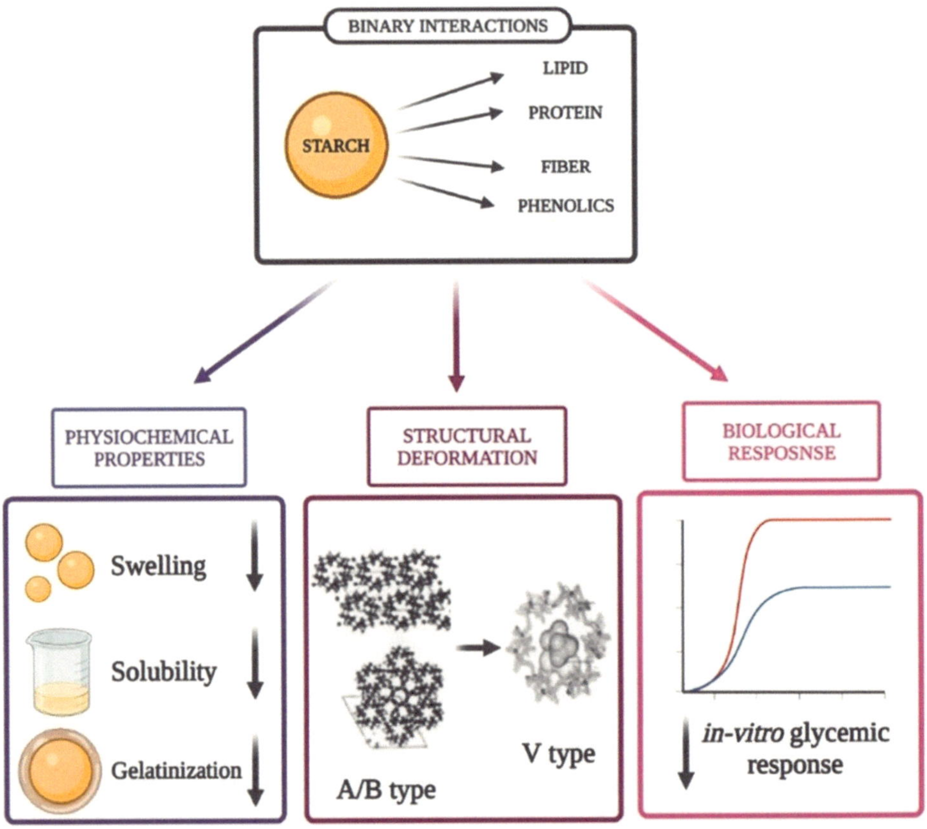

Figure 1.
Types of binary interactions and its effect on starch bioavailability. Binary interactions modulate physiochemical, structural, and biological attributes limiting starch digestibility as well as ultimate glycemic response.

such interactions on starch functional aspects like hydration, enzymatic cleavage, or enthalpy have been delineated using differential scanning calorimetry (DSC). Scanning electron microscopy (SEM) assisted in revealing the structural alterations associated with starch in the matrix after component depletion or addition. Rapid visco analysis (RVA) revealed that viscosity and pasting parameters were found inversely associated with in-vitro starch digestibility. The effect of matrix components in retaining the matrix, granule stability, preventing the expansion of granules as well as limiting the glycolytic enzyme attack has been endorsed using this technique [15, 16]. Other than affecting the swelling of starch granules by reducing the contact with carbolytic enzymes, the effect of such binary interactions in altering the molecular configuration (digestion sensitive A or B type to resistant V-type) of starch was envisioned and characterized using X-ray diffraction (XRD) and Fourier transform infrared microscope (FTIR).

Among the binary interactions, the most relevant in limiting the glycemic amplitude includes starch-lipid, starch-protein & starch-fiber dynamics.

1.1 Starch-lipid interactions

Even though well compartmentalized, starch and lipids do interact endogenously in real food systems. Lipid content ranges from 0.2–7% in cereals, with the least reported in rice and maximum reported in the case of oats & pearl millet [17]. A balanced distribution of neutral, glycol, and phospholipids along with free fatty acids have been reported in most of the food matrices, assist in energy as well as membrane structure & functions [17]. Curiosity towards food matrix interactions underlined a striking correlation between high lipid content [18] and low GR, which initiated binary (starch-lipid) interaction studies [7, 8]. Endogenous and exogenous lipid content have found to have low in-vitro starch digestibility along with superior resistant starch (RS) fraction. The effect of endogenous and exogenous lipid types have recently shown to have an effect in increasing starch-lipid complexation enriching RS content in red rice [9, 10]. Ye et al. [11] suggested that among lipids and proteins, starch digestibility is most affected by lipids as it affects swelling of granules, reduces the contact with carbolytic enzymes as well as alters the molecular structure from A-type into resistant V-type pattern. The long hydrophobic tail of lipid entering the cavity-like structure of amylose enables starch to form a stable complex, thereby hindering the accessibility of starch to enzyme attack [11]. In the case of mung bean flour, in-vitro starch digestibility and GI were increased significantly when endogenous lipids were removed [19]. Previous studies by Panyoo et al. [12], Krishnan et al. [9] have mentioned that stable starch-lipid complex results in a twist in digestibility phenotype into a digestion resistant fraction (RS-V), which caters to the gut microflora. As stated above, Copeland et al. [20]; Wang and Copeland [21] suggested this inclusion complex of starch-lipid also has an immense role in the food industry such as lowering solubility, swelling power, starch gelatinization, retrogradation, and enzyme action.

Starch-lipid complexes can exist inherently within the food matrix, or they may be produced by exogenous applications. A study by Obiro et al. [22]; reported that this complexation is mainly influenced by non-covalent interactions (hydrogen bonds, hydrophobic interactions, van der Waals interactions, and so on). Hydroxyl groups α-(1,4) are situated on the outer surface whereas methylene and oxygen groups present in the inner region of the complex strengthen the formation of starch-lipid complexes. Considering all positive impacts of starch-lipid complex, there are few factors (chain length of amylose, amylopectin, fatty acids, degree of unsaturation) that mostly govern the degree of complexation [10]. Various researchers stated that amylose acts as the primary constituent to interact with lipid molecules, while few reports supported the role of amylopectin chain length to form the complex [23, 24]. It has been reported from various studies that starch-lipid complexability has been increased with the longer chain length/degree of polymerization (DP) which highlights the formation of crystalline structure [25]. In addition to the effect exerted by the chain length of starch components, processing conditions like cooking also affect starch-lipid interaction. Kaur et al. [8] suggested amylose-lipid complexation enhanced with amylose chain length and increased with cooking time. Experiments highlighted the stability of starch-lipid inclusion complex formation mainly based on the types of fatty acids accommodated inside the helical cavity [26]. Different reports exist on the type of fatty acid for stable starch-lipid complexation. One school of thought suggests that the stability of the S-L complex could be enhanced by increasing the aliphatic chain length of fatty acids as well as melting temperature (from 8–10). On the other hand, another dimension highlights that smaller carbon chain length fatty acids might be more soluble into the aqueous solution and less stable also [27, 28]. Tufvesson et al. reported C14 as the most stable conformation than C16 or C18 while other explained C16 or C18 is better in the case of complexability [28]. Therefore, saturated fatty acid (SFA) with increased chain length can easily form a stable complex which further affects enzymatic accessibility due to resistance against carbolytic enzymes. Studies over decades highlighted that only SFA can be able to form a strong stable S-L complex with increasing chain length in a temperature-dependent manner whereas an inverse relationship has been found for unsaturated fatty acid (UFA) [26, 28]. A report from Zheng et al. [29] stated that chain length and degree of unsaturation have a role in the compact structure of starch-lipid formation. In addition to this, Kawai et al. [30] & Meng et al. [31] revealed starch-UFA complex showed resistance by formatting a stable complex to digestive enzyme action. The degree of complexability of FA in the case of maize starch ranged from 11.60–26.31% according to Sun et al. report [32]. Moreover, it has been explained from Sun et al. [32] RS is also enriched with the degree of unsaturation from 0 to 2%. In addition, thermal properties are also greatly affected by this S-L complex. Thermal complexes are mainly classified into two types of complexes as type I (90–115°C), type II (115–130°C) depending on the melting temperature. Studies from previous research have already highlighted that developed type II complex is more resistant to the digestive enzymes as compared to type I complex [33]. But Sun et al. [32] unraveled that maize starch-linoleic acid (MS-LOA) primarily formed as type I complex while maize starch-stearic acid (MS-SA) belonged to type I & type II complex. The reason behind this could be the large steric hindrance associated with LOA than SA which showed less accessibility of enzymes and inhibits ultimate glucose release. Cheng et al. [34] also used molecular dynamics to study amylose and linoleic acid structural analysis and conformational changes during complexation. On the continuation with Cheng et al., recently another research group of Schahl et al. [35] revealed the molecular structural complex using 13 NMR spectroscopy where they have taken quantum DFT approach affected by amylose size fragment and specific intramolecular hydrogen bonds. Hence, all the V-type complexes produced due to the addition of lipids act as a stable resistant structure against all digestive enzymes which further lowers glycemic response.

1.2 Starch-protein interactions

Proteins, mostly in the form of amino acids, and enzymes, are the predominant component in the food matrix, other than starch and fat [36]. Apart from the nutritional quality, proteins act as the major microstructural framework in a food matrix and hence also act as a physical barrier towards starch hydrolysis [37]. An interesting correlation among the reduction in insulinemic and glycemic responses by increasing the protein content in starchy crops led to the possibility of starch-protein interplay. Among the protein types, albumin, glutenins, and globulins aid in the gluing of protein bodies into a matrix enveloping the starch granules, which act as a barrier for starch digestion [38]. The existence of a protein barrier encircling the starch granule was validated using the pronase enzyme which dissociates the protein matrix and results in a considerable increase in-vitro starch digestibility [39]. Annor et al. [40] reported that the hypoglycemic characteristic of Kodo millet was related to the protein encircling the starch granules. Ren et al. [41] also reported that there was a fast increment in in-vivo GI and in-vitro starch digestibility of foxtail millet flour due to the lack of starch-protein complex after deproteination. Various studies have reported that the presence of gluten has an impact on the pace of starch digestion, resulting in reduced glycemic response [42, 43]. Gluten develops a visco-elastic and thick network that entraps starch granules, as well as a compact and stable structure that prevents starch granules from expanding and leaching during cooking, resulting in reduced accessibility of enzyme and slow-release properties [44]. To study the impact of protein removal from wheat products (bread) on blood glucose, healthy individuals were given meals of white bread prepared either from normal or gluten-free flour. It was observed that there was a considerable increase in blood glucose after consuming bread prepared from gluten-free flour. This led to an increase in digestion rate in-vitro and declined the starch mal-absorption in vivo as studied via breath-H_2 measurements, but this impact was not restored when the gluten was later added back to the gluten-free flour. The possible mechanism behind this may be all-purpose wheat flour is made up of granules with a starch core enveloped by a protein network that inhibits the hydrolysis rate in the small intestine lumen [45]. Recently, Lu et al. [46] revealed that in the small intestine, amino acids generated from enzymatic hydrolysis of rice protein inhibited the porcine pancreatic α-amylase activity. The protein content of rice flour was shown to be negatively associated with rapidly digestible starch (RDS) and slowly digestible starch (SDS), while positively with RS [47], on the other hand, the total protein content of rice grain was found to be inversely correlated with in-vivo GI [48].

Other than endogenous factors, processing (thermic/mechanical) has been found to have an effect in altering the level of interaction between protein-starch molecules, influencing the overall digestibility [49]. Pasini et al. [50] found that in-vitro digestion of wheat protein has been considerably reduced at elevated cooking temperatures (>180°C) due to the development of high molecular weight protein aggregates which are stabilized by strong irreversible linkages, distinct from hydrophobic and/or disulfide bonds that could be prevalent at low temperatures (100°C). Furthermore, it has been found that "appropriate" kneading/mixing promotes the development of a protein matrix (gluten) via disulfide linkages. Moreover, if extreme kneading/mixing is performed, the matrix loses strength as the linkages break and glutenin particles are fragmented into smaller fragments, which helps digestive enzymes access the starch and thus increases the starch digestibility [51].

Protein-enriched food formulations have also been found to impact the overall GI and thus assist in developing diabetic-friendly foods. Formulations based on proso millet starch and different protein mixtures (15% zein + 10% whey protein isolate + 15% soy protein isolate) reported that protein types reduced the RDS levels and enhanced RS levels from 4.49% to 11.73%. The blend comprising of corn starch (10%) and whey protein isolate had a considerably higher concentration of RS and low RDS as compared to pure corn starch. This could be due to the increased protein matrix enveloping starch networks, preventing amylolytic attack. When soy protein was added to maize starch, RDS was reduced while SDS and RS were increased [52]. The addition of 51% rice protein in wheat starch along with cellulose reduced RDS level, whereas the addition of protein from pea proteins (82%), maize (95%), soy (94%), and wheat (86%) did not affect RDS levels [13]. Bio-mimicking interactions in corn grains using microencapsulation of corn starch by zein protein have been reported with lowered starch digestibility [53]. Furthermore, starch coupled with amino acids or protein via the Maillard process has been demonstrated to limit the starch swelling, solubility as well as digestion rate [54], however, potential negative effects due to glycated product consumption must be examined in detail [55].

1.3 Starch-fiber interactions

Dietary fiber (DF), which consists primarily of non-starch polysaccharides found in plant cell walls, is an essential part of the food matrix [56]. DF types present in any food matrix are classified based on their water solubility and fermentability. Lignin, cellulose, and hemicelluloses are the major water-insoluble DF that get less fermented while the water-soluble DF includes pectin, mucilage, and gums and gets fermented properly in the small intestine [57]. Among the types, insoluble DF has been reported to be more useful in decreasing the GI as compared to their soluble fraction [58] as most of the common cereals contained a low level of naturally occurring soluble DF [59]. The endogenous fibers encircle the starch granules forming a starch-fiber network in the matrix, bio-mimicking an intact microstructure (plant cell/tissue) result in reduced enzyme accessibility and altered digestibility. Dense matrix composition in fiber content has been positively correlated to minimal postprandial GR after consumption in the case of barley, wheat, psyllium husk, and oats. This has been majorly attributed to the effect of insoluble DF in reducing starch bio accessibility as well as bioavailability [60, 61]. On the other hand, soluble DF like inulin has been found to form a protective barrier surrounding the starch granules, reducing starch swelling and release of amylose thus resulting in low viscosity values. This reduced the accessibility of starch-degrading enzymes that affect the in vivo starch digestibility and GI [62]. Among the studied types, β-glucan and guar gum have been reported to reduce the enzyme diffusion kinetics and thus the rate of carbohydrate digestion, eventually resulting in to slow down the gastric emptying and lower the liberation and absorption of glucose in the small intestine [63, 64]. The endogenous presence of β-glucans (native-form) in oats have been found to have an enveloping role towards starch and protein, thus reducing the enzyme accessibility, in turn, lowered starch digestibility and postprandial glycemia [65].

Endogenous presence, as well as exogenous addition of cellulose (insoluble fiber), has considerably reduced the α-amylase activity via mixed-type inhibition resulting in lowered in-vitro starch digestion [66]. The reduction of α-amylase activity was found to be positively linked with cellulose content, and α-amylase was found to be non-specifically linked on the surface of cellulose, reducing starch hydrolysis.

Interaction study between pectin and digestive enzyme (amyloglucosidase) showed a similar pattern, where pectin resulted in the conformational alteration in an enzyme that impeded substrate access and slower digestion rate of long amylopectin chains [67]. Luo and Zhang [68] aimed to mimic the microstructure of endosperm tissue by constructing a starch in a whole-grain-like structural form using calcium-induced alginate gelation in the presence of β-glucan and starch.

Processing strategies, as well as formulations with exogenous addition of fiber types, have also been found to reduce the in-vitro starch digestibility and GI of foods [69]. The addition of fibers like xanthan gum, glucomannan, and agar in rice lowered the starch digestibility in-vitro and in-vivo [70, 71]. However, no relationship was observed between native fiber content (0.5%) and in-vivo starch digestibility in rice, even though the fiber level was certainly too less to have any influence on starch digestion [48]. Reduction in blood glucose [72] and in-vitro starch digestibility [73] was observed in wheat products after adding β-glucan. β-Glucan has been assumed to improve viscosity, which could have lowered the rate of gastric emptying [72] and lowered the rate of diffusion of starch digestive enzymes. Vegetables like *Moringa oleifera* leaves and okra were found to reduce the glycemic response of various foods. When 10% okara was added to rice noodles, blood glucose levels significantly reduced [74]. Broccoli fiber addition in a potato diet has proved to assist in decreasing the GI by increasing RS content [75]. However, a 30% decrease in GI in the same sample was observed when studied in-vivo. Hardacre et al. [76] showed that fibers with comparable viscosities resulted in variation of in- vitro starch digestion and hypothesized that few fibers may inhibit certain enzymes in a non-competitive manner as a chemical barrier. An interesting observation reported by Sciarini et al. [69] was that, the addition of up to 5% soluble (Inulin) and insoluble fibers (oat fiber and type IV RS) enhanced the starch digestibility in GF bread, while further increase resulted in a reduction in starch digestibility. The initial observed increase could be due to the altered bread crumb structure while a higher percentage of fiber could have established a staple starch-fiber network and thus reducing the digestibility. The impact of RS addition in pasta structure on native starch digestibility was studied by Gelencser et al. [77]. They reported that kinetic characteristics were not considerably variable between the control and RS-added samples, whereas the starch digestibility was considerably low in the RS-added samples, signifying a decrease in absolute glucose release during amylolysis. The addition of RSII (7.5%) and RSIV (10%) in pasta showed a decrease in in-vitro starch digestibility and GI [78]. Furthermore, larger DF concentrations also might play role in confining the starch inside the pores, preventing its hydrolysis. Mkandawire et al. [79] discovered that adding up to 50% cellulose (w/w starch) to sorghum flours had no significant impact on the RS level. DF, on the other hand, enhanced the solution viscosity in-vivo, and therefore may slow starch degradation by restricting enzyme mobility, as gums do, and hence slowing digestion rate in total [63].

In this direction, several animal studies have been carried out to study the effect of adding fiber on starch bioavailability or glucose release. The supplement of insoluble cereal DF from oat leads to enhanced insulin sensitivity in obese mice [80]. Further studies conducted by Weickert et al. [81] revealed that oat DF and purified wheat can enhance the postprandial insulin secretion hormones which further improved the postprandial carbohydrate metabolism. The high level (500 mg/kg body weight) of oat β-glucan or 4% barley β-glucan resulted in considerable enhancement of insulin resistance in insulin-resistant mice model and the impact was concentration-dependent [82, 83]. β-Glucan was found to inhibit the intestinal disaccharides' activities in-vitro and in-vivo, which led to slow starch digestion rate [84]. In the diabetic mice model, β-glucan considerably repaired and increased the integrity of pancreatic islet β-cell and tissue structures [85]. Overall, the type and concentration of fiber have a customized effect on the food matrix. Comprehensive list of various food components added to starch and their effect on starch digestibility is tabulated in **Table 1**.

S. no.	Food component added	Added to starch	Impact on starch digestibility	Reference
	Lipid			
1.	Cooking fats (ghee, coconut oil, virgin coconut oil, rice bran oil) (2.5%)	Rice starch (white, black, red)	↓	[9]
2.	Linoleic acid (0.75%)	Arrowhead tubers starch	↓	[86]
3.	Ascorbyl palmitate (10%)	High amylose maize starch, potato starch	↓	[87]
4.	Trans-oleic acid, cis-oleic acid, cis linoleic acid (1%, 3%, 5%)	Rice starch	↓	[87]
5.	Palmitic acid (0.5%)	Waxy rice starch	↓	[88]
6.	Oleic acid (1%, 2%, 3%) & linoleic acid (2%, 4%, 6%)	Rice starch	↓	[89]
7.	Linoleic acid, monomyristyl glycerol (0%, 0.5%, 1%, 1.5%, 2%, 3%, 5%)	Maize starch	↓	[90]
8.	Oleic acid (4%)	Native rice starch	↓	[91]
9.	Palm oil (5%, 10%)	Arrowroot starches	↓	[92]
10.	Lauric acid (1.5%)	Wheat starch	↓	[93]
11.	Oleic acid (0.05%)	Native potato starch	↓	[94]
12.	Dodecanoic acid, tetradecanoic acid, octadecanoic acid (1%, 3%, 5%)	Native rice starch	↓	[95]
13.	Decanoic acid, palmitic acid (10%)	Native maize starch	↓	[32]
14.	Lauric acid, stearic acid and glycerides (glycerol monolaurate, glycerol monostearate) (0.18%)	Wheat starch	↓	[93]
15.	Palmitic acid (0.3%)	Maize starch	↓	[1]
	Protein			
16.	Rice globulin (2%)	Rice starch	↓	[15]
17.	Common bean (15%, 30%, 45%)	Wheat semolina	↓	[96]
18.	Gluten (2%)	Wheat flour	↓	[97]
19.	Beans (10%, 20%, 30%)	Semolina flour	↓	[98]
20.	White beans (15%, 30%, 45%)	Rice flour	↓	[99]
21.	Rice protein (51%)	Wheat starch	↓	[52]

S. no.	Food component added	Added to starch	Impact on starch digestibility	Reference
22.	Alfalfa seed (15%, 30%, 45%)	Rice flour	↓	[100]
23.	Hydrolyzed protein (12%)	Wheat flour	↓	[13]
24.	Gluten (20%)	Wheat starch	↓	[101]
25.	Whey protein isolate (2.5%, 4.5%, 10%)	Native corn starch	↓	[102]
26.	Rice globulin (2.5%)	Rice flour	No effect on digestibility	[103]
27.	Soybean peptide (5%, 10%, 15%)	Corn or potato starch	↓	[104]
28.	Chickpea protein (8%)	Rice flour	↓	[105]
	Fiber			
29.	Oat fiber (10%)	GF bread	↓	[70]
30.	Oat fiber (>5%)	Pasta	↓	[106]
31.	Inulin (12%)	GF bread	↓	[107]
32.	Cellulose (50%)	Potato starch	↓	[79]
33.	Cellulose (9–83%)	Maize starch	↓	[66]
34.	*Lentinus* β-glucan (20%)	Wheat flour	↓	[73]
35.	Glucomannan (0.1–0.2%)	Rice starch	↓	[70]
36.	Xanthun gum (0.4%)	Rice starch	↓	[71]
37.	RSIV (Novelose 480) (10%)	Pasta	↓	[108]
38.	RSIV (Novelose 480) (10%)	GF bread	↓	[69]
39.	RSII (Native HA maize starch) (20%)	GF bread	↓	[109]
40.	RSII (Hi-maize^{Tm} 260) (20%)	Pasta	↓	[106]
41.	RSII (Fibersym^{Tm}70) (20%)	Pasta	↓	[77]
42.	Okara (10%)	Rice noodles	↓	[74]

Table 1.
List of various food components added to starch and their effect on starch digestibility.

2. Conclusion

Binary interactions among the nutrient types and starch mediate the glycemic amplitude of real food systems. Among the binary interactions (starch-lipid, starch-protein, starch-fiber), the role has been extensively characterized in limiting the enzyme penetrance, altering the molecular configuration, starch digestibility, and thus in turn GR. Understanding such binary interactions, not only shares a logical explanation for the low GI of whole-grain foods but also the immense role of such cereals in diabetic-friendly foods. Even though the existing rationale supports the fact that multiple food matrix interaction studies at a time are difficult, it's indeed vital to study ternary (three-way) and quaternary (four-way) interactions and their role in limiting the glycemic response. Finally, it's important to keep in mind that altering starch's nutritional qualities can also change its desired physicochemical and sensory qualities, affecting food quality that should be considered while developing novel foods.

Conflict of interest

The authors declare no conflict of interest.

Author details

Veda Krishnan[1*], Monika Awana[1], Debarati Mondal[1], Piyush Verma[2], Archana Singh[1] and Shelly Praveen*

1 Division of Biochemistry, ICAR-Indian Agricultural Research Institute (IARI), New Delhi, India

2 School of Pharmaceutical Science and Technology, Sardar Bhagwan Singh University, Dehradun, Uttarakhand, India

*Address all correspondence to: veda.krishnan@icar.gov.in, vedabiochem@gmail.com and shelly.parveen@icar.gov.in

IntechOpen

© 2022 The Author(s). Licensee IntechOpen. This chapter is distributed under the terms of the Creative Commons Attribution License (http://creativecommons.org/licenses/by/3.0), which permits unrestricted use, distribution, and reproduction in any medium, provided the original work is properly cited.

References

[1] Li Q, Shi S, Du SK, Dong Y, Yu X. Starch–palmitic acid complex formation and characterization at different frying temperature and treatment time. LWT. 2021;**136**:110328. DOI: 10.1016/j.lwt.2020.110328

[2] Wang J, Jiang X, Zheng B, Zhang Y. Structural and physicochemical properties of lotus seed starch-chlorogenic acid complexes prepared by microwave irradiation. Journal of Food Science and Technology. 2021;**58**:4157-4166. DOI: 10.1007/s13197-020-04881-w

[3] Krishnan V, Awana M, Singh A, Goswami S, Vinutha T, Kumar RR, et al. Starch molecular configuration and starch-sugar homeostasis: Key determinants of sweet sensory perception and starch hydrolysis in pearl millet (*Pennisetum glaucum*). International Journal of Biological Macromolecules. 2021;**183**:1087-1095. DOI: 10.1016/j.ijbiomac.2021.05.004

[4] Krishnan V, Rani R, Awana M, Pitale D, Kulshreshta A, Sharma S, et al. Role of nutraceutical starch and proanthocyanidins of pigmented rice in regulating hyperglycemia: Enzyme inhibition, enhanced glucose uptake and hepatic glucose homeostasis using in vitro model. Food Chemistry. 2021;**335**:127505. DOI: 10.1016/j.foodchem.2020.127505

[5] Krishnan V, Mondal D, Thomas B, Singh A, Praveen S. Starch-lipid interaction alters the molecular structure and ultimate starch bioavailability: A comprehensive review. International Journal of Biological Macromolecules. 2021;**182**:626-638. DOI: 10.1016/j.ijbiomac.2021.04.030

[6] Krishnan V, Awana M, Rani AR, Bansal N, Bollinedi H, Srivastava S, et al. Quality matrix reveals the potential of Chak-hao as a nutritional supplement: A comparative study of matrix components, antioxidants and physicochemical attributes. Journal of Food Measurement and Characterization. 2021;**15**:826-840. DOI: 10.1007/s11694-020-00677-w

[7] Henry CJK, Lightowler HJ, Newens KJ, Pata N. The influence of adding fats of varying saturation on the glycaemic response of white bread. International Journal of Food Sciences and Nutrition. 2008;**59**:61-69. DOI: 10.1080/09637480701664183

[8] Kaur B, Ranawana V, Teh AL, Henry CJK. The glycemic potential of white and red rice affected by oil type and time of addition. Journal of Food Science. 2015;**80**:H2316-H2321. DOI: 10.1111/1750-3841.13070

[9] Krishnan V, Mondal D, Bollinedi H, Srivastava S, Ramesh SV, Madhavan L, et al. Cooking fat types alter the inherent glycaemic response of niche rice varieties through resistant starch (RS) formation. International Journal of Biological Macromolecules. 2020;**162**:1668-1681. DOI: 10.1016/j.ijbiomac.2020.07.265

[10] Krishnan V, Awana M, Samota MK, Warwate SI, Kulshreshtha A, Ray M, et al. Pullulanase activity: A novel indicator of inherent resistant starch in rice (*Oryza sativa* L). International Journal of Biological Macromolecules. 2020;**152**:1213-1223. DOI: 10.1016/j.ijbiomac.2019.10.218

[11] Ye J, Hu X, Luo S, McClements DJ, Liang L, Liu C. Effect of endogenous proteins and lipids on starch digestibility in rice flour. Food Research International. 2018;**106**:404-409. DOI: 10.1016/j.foodres.2018.01.008

[12] Panyoo AE, Emmambux MN. Amylose-lipid complex production and potential health benefits: A mini-review. Starch-Starke. 2017;**69**:1600203. DOI: 10.1002/star.201600203

[13] Lopez-Baron N, Gu Y, Vasanthan T, Hoover R. Plant proteins mitigate in vitro wheat starch digestibility. Food Hydrocolloids. 2017;**69**:19-27. DOI: 10.1016/j.foodhyd.2017.01.015

[14] Jin Z, Bai F, Chen Y, Bai B. Interactions between protein, lipid and starch in foxtail millet flour affect the in vitro digestion of starch. CyTA-Journal of Food. 2019;**17**:640-647. DOI: 10.1080/19476337.2019.1628107

[15] Baxter G, Zhao J, Blanchard C. Albumi significantly affects pasting and textural characteristics of rice flour. Cereal Chemistry. 2010;**87**:250-255. DOI: 10.1094/CCHEM-87-3-0250

[16] Chung HJ, Liu Q, Lee L, Wei D. Relationship between the structure, physicochemical properties and in vitro digestibility of rice starches with different amylose contents. Food Hydrocolloids. 2011;**25**:968-975. DOI: 10.1016/j.foodhyd.2010.09.011

[17] Price PB, Parsons JG. Lipids of seven cereal grains. Journal of the American Oil Chemists Society. 1975;**52**(12):490-493. DOI: 10.1007/BF02640738

[18] Li Q, Wu QY, Jiang W, Qian JY, Zhang L, Wu M, et al. Effect of pulsed electric field on structural properties and digestibility of starches with different crystalline type in solid state. Carbohydrate Polymers. 2019;**207**:362-370. DOI: 10.1016/j.carbpol.2018.12.001

[19] Hou D, Zhao Q, Yousaf L, Xue Y, Shen Q. In vitro starch digestibility and estimated glycemic index of mung bean (*Vigna radiata* L.) as affected by endogenous proteins and lipids, and exogenous heat-processing methods. Plant Foods for Human Nutrition. 2020;**75**:547-552. DOI: 10.1007/s11130-020-00845-9

[20] Copeland L, Blazek J, Salman H, Tang MC. Form and functionality of starch. Food Hydrocolloids. 2009;**23**:1527-1515. DOI: 10.1016/j.foodhyd.2008.09.016

[21] Wang S, Copeland L. Molecular disassembly of starch granules during gelatinization and its effect on starch digestibility: A review. Food and Function. 2013;**4**(11):1564-1580. DOI: 10.1039/D1FO00729G

[22] Obiro WC, Sinha Ray S, Emmambux MN. V-amylose structural characteristics, methods of preparation, significance, and potential applications. Food Reviews International. 2012;**28**(4):412-438. DOI: 10.1080/87559129.2012.660718

[23] Wang S, Wang J, Yu J, Wang S. Effect of fatty acids on functional properties of normal wheat and waxy wheat starches: A structural basis. Food Chemistry. 2016;**190**:285-292. DOI: 10.1016/j.foodchem.2015.05.086

[24] Eliasson AC. Interactions between starch and lipids studied by DSC. Thermochimica Acta. 1994;**246**:343-356. DOI: 10.1016/0040-6031(94)80101-0

[25] Gelders GG, Vanderstukken TC, Goesaert H, Delcour JA. Amylose–lipid complexation: A new fractionation method. Carbohydrate Polymers. 2004;**56**:447-458. DOI: 10.1016/j.carbpol.2004.03.012

[26] Eliasson AC, Krog N. Physical properties of amylose-monoglyceride complexes. Journal of Cereal Science.

1985;**3**:239-248. DOI: 10.1016/S0733-5210(85)80017-5

[27] Qin R, Yu J, Li Y, Copeland L, Wang S, Wang S. Structural changes of starch–lipid complexes during postprocessing and their effect on in vitro enzymatic digestibility. Journal of Agricultural and Food Chemistry. 2019;**67**(5):1530-1536. DOI: 10.1021/acs.jafc.8b06371

[28] Tufvesson F, Wahlgren M, Eliasson AC. Formation of amylose-lipid complexes and effects of temperature treatment. Part 2. Fatty acids. Starch-Starke. 2003;**55**(3-4):138-149. DOI: 10.1002/star.200390028

[29] Zheng M, Chao C, Yu J, Copeland L, Wang S, Wang S. Effects of chain length and degree of unsaturation of fatty acids on structure and in vitro digestibility of starch–protein–fatty acid complexes. Journal of Agricultural and Food Chemistry. 2018;**66**(8):1872-1880. DOI: 10.1111/ijfs.14347

[30] Kawai K, Takato S, Sasaki T, Kajiwara K. Complex formation, thermal properties, and in-vitro digestibility of gelatinized potato starch–fatty acid mixtures. Food Hydrocolloids. 2012;**27**:228-234. DOI: 10.1016/j.foodhyd.2011.07.003

[31] Meng S, Ma Y, Sun DW, Wang L, Liu T. Properties of starch-palmitic acid complexes prepared by high pressure homogenization. Journal of Cereal Science. 2014;**59**(1):25-32. DOI: 10.1016/j.jcs.2013.10.012

[32] Sun S, Jin Y, Hong Y, Gu Z, Cheng L, Li Z, et al. Effects of fatty acids with various chain lengths and degrees of unsaturation on the structure, physicochemical properties and digestibility of maize starch-fatty acid complexes. Food Hydrocolloids. 2021;**110**:106224. DOI: 10.1016/j.foodhyd.2020.106224

[33] Liu P, Gao W, Zhang X, Wu Z, Yu B, Cui B. Physicochemical properties of pea starch-lauric acid complex modified by maltogenic amylase and pullulanase. Carbohydrate Polymers. 2020; **242**:116332. DOI: 10.1016/j.carbpol.2020.116332

[34] Cheng L, Feng T, Zhang B, Zhu X, Hamaker B, Zhang H, et al. A molecular dynamics simulation study on the conformational stability of amylose-linoleic acid complex in water. Carbohydrate Polymers. 2018;**196**:56-65. DOI: 10.1016/j.carbpol.2018.04.102

[35] Schahl A. Interactions polysaccharides-lipides: étude théorique et expérimentale combinant calculs de dynamique moléculaire, calculs quantiques de spectres RMN 13C et RMN 13C à l'état solide. Science des matériaux [cond-mat.mtrl-sci]. Français: Université Paul Sabatier - Toulouse III, 2020. ⟨NNT: 2020TOU30178⟩

[36] Lal MK, Singh B, Sharma S, Singh MP, Kumar A. Glycemic index of starchy crops and factors affecting its digestibility: A review. Trends in Food Science and Technology. 2021;**111**:741-755. DOI: 10.1016/j.tifs.2021.02.067

[37] Svihus B, Uhlen AK, Harstad OM. Effect of starch granule structure, associated components and processing on nutritive value of cereal starch: A review. Animal Feed Science and Technology. 2005;**122**(3-4):303-320. DOI: 10.1016/j.anifeedsci.2005.02.025

[38] Hamaker BR, Bugusu BA. Overview: Sorghum proteins and food quality. Pretoria, South Africa. In: Workshop on the proteins of sorghum and millets: Enhancing nutritional and functional properties for Africa [Internet]. 2003.

Available from: http://www.afripro.org.uk/papers/Paper08Hamaker.pdf

[39] Rooney LW, Pflugfelder RL. Factors affecting starch digestibility with special emphasis on sorghum and corn. Journal of Animal Science. 1986;**63**(5):1607-1623. DOI: 10.2527/jas1986.6351607x

[40] Annor GA, Marcone M, Bertoft E, Seetharaman K. In vitro starch digestibility and expected glycemic index of Kodo millet (*Paspalum scrobiculatum*) as affected by starch-protein-lipid interactions. Cereal Chemistry Journal. 2013;**90**:211-217. DOI: 10.1094/CCHEM-06-12-0074-R

[41] Ren X, Chen J, Wang C, Molla MM, Diao X, Shen Q. In vitro starch digestibility, degree of gelatinization and estimated glycemic index of foxtail millet-derived products: Effect of freezing and frozen storage. Journal of Cereal Science. 2016;**69**:166-173. DOI: 10.1016/j.jcs.2016.03.007

[42] Granfeldt Y, Bjorck I. Glycemic response to starch in pasta: A study of mechanisms of limited enzyme availability. Journal of Cereal Science. 1991;**14**:47-61. DOI: 10.1016/S0733-5210(09)80017-9

[43] Kim EHJ, Petrie JR, Motoi L, Morgenstern MP, Sutton KH, Mishra S, et al. Effect of structural and physicochemical characteristics of the protein matrix in pasta on in vitro starch digestibility. Food Biophysics. 2008;**3**:229-234. DOI: 10.1007/s11483-008-9066-7

[44] Zou W, Sissons M, Warren FJ, Gidley MJ, Gilbert RG. Compact structure and proteins of pasta retard in vitro digestive evolution of branched starch molecular structure. Carbohydrate Polymers. 2016;**152**:441-449. DOI: 10.1016/j.carbpol.2016.06.016

[45] Jenkins DJ, Thorne MJ, Wolever TM, Jenkins AL, Rao AV, Thompson LU. The effect of starch-protein interaction in wheat on the glycemic response and rate of in vitro digestion. The American Journal of Clinical Nutrition. 1987; **45**:946-951. DOI: 10.1093/ajcn/45.5.946

[46] Lu X, Chang R, Lu H, Ma R, Qiu L, Tian Y. Effect of amino acids composing rice protein on rice starch digestibility. LWT. 2021;**146**:111417. DOI: 10.1016/j.lwt.2021.111417

[47] Zhu LJ, Liu QQ, Wilson JD, Gu MH, Shi YC. Digestibility and physicochemical properties of rice (*Oryza sativa* L.) flours and starches differing in amylose content. Carbohydrate Polymers. 2011;**86**(4): 1751-1759. DOI: 10.1016/j.carbpol.2011.07.017

[48] Zarati M, Pirali M, Mirmiran P, Nouri N, Nakhoda K, Najafi H, et al. Glycemic index of various brands of rice in healthy individuals. International Journal of Endocrinology and Metabolism. 2008;**6**(4):200-204. Available from: https://www.sid.ir/en/journal/ViewPaper.aspx?id=134381

[49] Parada J, Santos JL. Interactions between starch, lipids, and proteins in foods: Microstructure control for glycemic response modulation. Critical Reviews in Food Science and Nutrition. 2016;**56**(14):2362-2369. DOI: 10.1080/10408398.2013.840260

[50] Pasini G, Simonato B, Giannattasio M, Peruffo AD, Curioni A. Modifications of wheat flour proteins during in vitro digestion of bread dough, crumb, and crust: An electrophoretic and immunological study. Journal of Agricultural and Food Chemistry. 2001;**49**(5):2254-2261. DOI: 10.1021/jf0014260

[51] Parada J, Aguilera JM. Microstructure, mechanical properties, and starch digestibility of a cooked dough made with potato starch and wheat gluten. LWT-Food Science and Technology. 2011;**44**(8):1739-1744. DOI: 10.1016/j.lwt.2011.03.012

[52] Chen X, He XW, Zhang B, Fu X, Jane JL, Huang Q. Effects of adding corn oil and soy protein to corn starch on the physicochemical and digestive properties of the starch. International Journal of Biological Macromolecules. 2017;**104**:481-486. DOI: 10.1016/j.ijbiomac.2017.06.024

[53] Xu H, Zhang G. Slow digestion property of microencapsulated normal corn starch. Journal of Cereal Science. 2014;**60**(1):99-104. DOI: 10.1016/j.jcs.2014.01.021

[54] Yang W, Hattori M, Kawaguchi T, Takahashi K. Properties of starches conjugated with lysine and poly(lysine) by the Maillard reaction. Journal of Agricultural and Food Chemistry. 1998;**46**(2):442-445. DOI: 10.1021/jf970515i

[55] Miao M, Hamaker BR. Food matrix effects for modulating starch bioavailability. Annual Review of Food Science and Technology. 2021;**12**:169-191. DOI: 10.1146/annurev-food-070620-013937

[56] Lal MK, Kumar A, Kumar A, Raigond P, Oko AO, Thakur N, et al. Dietary fibres in po66to. In: Potato. Singapore: Springer; 2020. pp. 37-50

[57] Xie F, Li M, Lan X, Zhang W, Gong S, Wu J, et al. Modification of dietary fibers from purple-fleshed potatoes (heimeiren) with high hydrostatic pressure and high pressure homogenization processing: A comparative study. Innovative Food Science and Emerging Technologies. 2017;**42**:157-164. DOI: 10.1016/j.ifset.2017.05.012

[58] Krishnan S, Rosenberg L, Singer M, Hu FB, Djousse L, Cupples LA, et al. Glycemic index, glycemic load, and cereal fiber intake and risk of type 2 diabetes in US black women. Archives of Internal Medicine. 2007;**167**:2304-2309. DOI: 10.1001/archinte.167.21.2304

[59] Aldughpassi A, Abdel-Aal ESM, Wolever TM. Barley cultivar, kernel composition, and processing affect the glycemic index. The Journal of Nutrition. 2012;**142**:1666-1671. DOI: 10.3945/jn.112.161372

[60] Ou S, Kwok KC, Li Y, Fu L. In vitro study of possible role of dietary fiber in lowering postprandial serum glucose. Journal of Agricultural and Food Chemistry. 2001;**49**(2):1026-1029. DOI: 10.1021/jf000574n

[61] Ahmed F, Sairam S, Urooj A. In vitro hypoglycemic effects of selected dietary fiber sources. Journal of Food Science and Technology. 2011;**48**:285-289. DOI: 10.1007/s13197-010-0153-7

[62] Vazquez-Gutierrez JL, Johansson D, Langton M. Effects of added inulin and wheat gluten on structure of rye porridge. LWT-Food Science and Technology. 2016;**66**:211-216. DOI: 10.1016/j.lwt.2015.10.034

[63] Singh J, Dartois A, Kaur L. Starch digestibility in food matrix: A review. Trends in Food Science and Technology. 2010;**21**(4):168-180. DOI: 10.1016/j.tifs.2009.12.001

[64] Regand A, Chowdhury Z, Tosh SM, Wolever TM, Wood P. The molecular weight, solubility and viscosity of oat beta-glucan affect human glycemic response by modifying starch

digestibility. Food Chemistry. 2011;**129**(2):297-304. DOI: 10.1016/j.foodchem.2011.04.053

[65] Zhang J, Luo K, Zhang G. Impact of native form oat β-glucan on starch digestion and postprandial glycemia. Journal of Cereal Science. 2017;**73**:84-90. DOI: 10.1016/j.jcs.2016.11.013

[66] Dhital S, Gidley MJ, Warren FJ. Inhibition of α-amylase activity by cellulose: Kinetic analysis and nutritional implications. Carbohydrate Polymers. 2015;**123**:305-312. DOI: 10.1016/j.carbpol.2015.01.039

[67] Bai Y, Wu P, Wang K, Li C, Li E, Gilbert RG. Effects of pectin on molecular structural changes in starch during digestion. Food Hydrocolloids. 2017;**69**:10-18. DOI: 10.1016/j.foodhyd.2017.01.021

[68] Luo K, Zhang G. Nutritional property of starch in a whole-grain-like structural form. Journal of Cereal Science. 2018;**79**:113-117. DOI: 10.1016/j.jcs.2017.09.006

[69] Sciarini LS, Bustos MC, Vignola MB, Paesani C, Salinas CN, Perez GT. A study on fibre addition to gluten free bread: Its effects on bread quality and in vitro digestibility. Journal of Food Science and Technology. 2017;**54**(1):244-252. DOI: 10.1007/s13197-016-2456-9

[70] Sasaki T, Kohyama K. Effect of non-starch polysaccharides on the in vitro digestibility and rheological properties of rice starch gel. Food Chemistry. 2011;**127**(2):541-546. DOI: 10.1016/j.foodchem.2011.01.038

[71] Oh IK, Bae IY, Lee HG. Hypoglycemic effect of dry heat treated starch with xanthan: An in vitro and in vivo comparative study. Starch-Stärke. 2018;**70**(9-10):1800088. DOI: 10.1002/star.201800088

[72] Cavallero A, Empilli S, Brighenti F, Stanca AM. High (1→3, 1→4)-β-glucan barley fractions in bread making and their effects on human glycemic response. Journal of Cereal Science. 2002;**36**:59-66. DOI: 10.1006/jcrs.2002.0454

[73] Zhuang H, Chen Z, Feng T, Yang Y, Zhang J, Liu G, et al. Characterization of *Lentinus edodes* β-glucan influencing the in vitro starch digestibility of wheat starch gel. Food Chemistry. 2017;**224**:294-301. DOI: 10.1016/j.foodchem.2016.12.087

[74] Kang MJ, Bae IY, Lee HG. Rice noodle enriched with okara: Cooking property, texture, and in vitro starch digestibility. Food Bioscience. 2018;**22**:178-183. DOI: 10.1016/j.fbio.2018.02.008

[75] Ballance S, Knutsen SH, Fosvold W, Wickham M, Trenado CDT, Monro J. Glyceamic and insulinaemic response to mashed potato alone, or with broccoli, broccoli fibre or cellulose in healthy adults. European Journal of Nutrition. 2018;**57**:199-207. DOI: 10.1007/s00394-016-1309-7

[76] Hardacre AK, Yap SY, Lentle RG, Monro JA. The effect of fibre and gelatinised starch type on amylolysis and apparent viscosity during in vitro digestion at a physiological shear rate. Carbohydrate Polymers. 2015;**123**:80-88. DOI: 10.1016/j.carbpol.2015.01.013

[77] Gelencser T, Gal V, Salgo A. Effects of applied process on the in vitro digestibility and resistant starch content of pasta products. Food and Bioprocess Technology. 2010;**3**:491-497. DOI: 10.1007/s11947-008-0105-7

[78] Li Y, Lv J, Wang L, Zhu Y, Shen R. Effects of millet bran dietary fiber and millet flour on dough development, steamed bread quality, and digestion in

vitro. Applied Sciences. 2020;**10**:912. DOI: 10.3390/app10030912

[79] Mkandawire NL, Kaufman RC, Bean SR, Weller CL, Jackson DS, Rose DJ. Effects of sorghum (*Sorghum bicolor* (L.) Moench) tannins on α-amylase activity and in vitro digestibility of starch in raw and processed flours. Journal of Agricultural and Food Chemistry. 2013;**61**(18):4448-4454. DOI: 10.1021/jf400464j

[80] Isken F, Klaus S, Osterhoff M, Pfeiffer AF, Weickert MO. Effects of long-term soluble vs. insoluble dietary fiber intake on high-fat diet-induced obesity in C57BL/6J mice. The Journal of Nutritional Biochemistry. 2010;**21**:278-284. DOI: 10.1016/j.jnutbio.2008.12.012

[81] Weickert MO, Mohlig M, Koebnick C, Holst JJ, Namsolleck P, Ristow M, et al. Impact of cereal fibre on glucose regulating factors. Diabetologia. 2005;**48**(11):2343-2353. DOI: 10.1007/s00125-005-1941-x

[82] Zheng J, Shen N, Wang S, Zhao G. Oat beta-glucan ameliorates insulin resistance in mice fed on high-fat and high-fructose diet. Food and Nutrition Research. 2013;**57**(1):22754. DOI: 10.3402/fnr.v57i0.22754

[83] Choi JS, Kim H, Jung MH, Hong S, Song J. Consumption of barley β-glucan ameliorates fatty liver and insulin resistance in mice fed a high-fat diet. Molecular Nutrition and Food Research. 2010;**54**:1004-1013. DOI: 10.1002/mnfr.200900127

[84] Dong J, Cai F, Shen R, Liu Y. Hypoglycaemic effects and inhibitory effects on intestinal disaccharidases of oat beta-glucan in streptozotocin-induced diabetic mice. Food Chemistry. 2011;**129**:1066-1071. DOI: 10.1016/j.foodchem.2011.05.076

[85] Liu M, Zhang Y, Zhang H, Hu B, Wang L, Qian H, et al. The anti-diabetic activity of oat β-d-glucan in streptozotocin–nicotinamide induced diabetic mice. International Journal of Biological Macromolecules. 2016;**91**:1170-1176. DOI: 10.1016/j.ijbiomac.2016.06.083

[86] Raza H, Ameer K, Ren X, Liang Q, Chen X, Chen H, et al. Physicochemical properties and digestion mechanism of starch-linoleic acid complex induced by multi-frequency power ultrasound. Food Chemistry. 2021;**364**:130392. DOI: 10.1016/j.foodchem.2021.130392

[87] Guo T, Hou H, Liu Y, Chen L, Zheng B. In vitro digestibility and structural control of rice starch-unsaturated fatty acid complexes by high-pressure homogenization. Carbohydrate Polymers. 2021;**256**:117607. DOI: 10.1016/j.carbpol.2020.117607

[88] Luo S, Zeng Z, Mei Y, Huang K, Wu J, Liu C, et al. Improving ordered arrangement of the short-chain amylose-lipid complex by narrowing molecular weight distribution of short-chain amylose. Carbohydrate Polymers. 2020;**240**:116359. DOI: 10.1016/j.carbpol.2020.116359

[89] He H, Zheng B, Wang H, Li X, Chen L. Insights into the multi-scale structure and in vitro digestibility changes of rice starch-oleic acid/linoleic acid complex induced by heat-moisture treatment. Food Research International. 2020;**137**:109612. DOI: 10.1016/j.foodres.2020.109612

[90] Chao C, Huang S, Yu J, Copeland L, Wang S, Wang S. Molecular mechanisms underlying the formation of starch-lipid complexes during simulated food processing: A dynamic structural analysis. Carbohydrate Polymers. 2020;**244**:116464. DOI: 10.1016/j.carbpol.2020.116464

[91] Zheng B, Wang T, Wang H, Chen L, Zhou Z. Studies on nutritional intervention of rice starch-oleic acid complex (resistant starch type V) in rats fed by high-fat diet. Carbohydrate Polymers. 2020;**246**:116637. DOI: 10.1016/j.carbpol.2020.116637

[92] Handarini K, SaumanHamdani J, Cahyana Y, SitiSetiasih I. Functional and pasting properties of a starch–lipid complex formed with gaseous ozone and palm oil. International Journal of Food Properties. 2020;**23**(1):1361-1372. DOI: 10.1080/10942912.2020.1801723

[93] Kang X, Yu B, Zhang H, Sui J, Guo L, Abd El-Aty AM, et al. The formation and in vitro enzymatic digestibility of starch-lipid complexes in steamed bread free from and supplemented with different fatty acids: Effect on textural and retrogradation properties during storage. International Journal of Biological Macromolecules. 2021;**166**:1210-1219. DOI: 10.1016/j.ijbiomac.2020.11.003

[94] Krolikowska K, Pietrzyk S, Labanowska M, Kurdziel M, Pajak P. The influence of acid hydrolysis on physicochemical properties of starch-oleic acid mixtures and generation of radicals. Food Hydrocolloids. 2021;**118**:106780. DOI: 10.1016/j.foodhyd.2021.106780

[95] Cui J, Zheng B, Liu Y, Chen L, Li B, Li L. Insights into the effect of structural alternations on the digestibility of rice starch-fatty acid complexes prepared by high-pressure homogenization. LWT. 2021;**136**:110294. DOI: 10.1016/j.lwt.2020.110294

[96] Gallegos-Infante JA, Bello-Perez LA, Rocha-Guzman NE, Gonzalez-Laredo RF, Avila-Ontiveros M. Effect of the addition of common bean (*Phaseolus vulgaris* L.) flour on the in vitro digestibility of starch and undigestible carbohydrates in spaghetti. Journal of Food Science. 2010;**75**(5): H151-H156. DOI: 10.1111/j.1750-3841.2010.01621.x

[97] Aravind N, Sissons M, Fellows C. Can variation in durum wheat pasta protein and starch composition affect in vitro starch hydrolysis? Food Chemistry. 2011;**124**(3):816-821. DOI: 10.1016/j.foodchem.2010.07.002

[98] Kumar SB, Prabhasankar P. A study on starch profile of rajma bean (*Phaseolus vulgaris*) incorporated noodle dough and its functional characteristics. Food Chemistry. 2015;**180**:124-132. DOI: 10.1016/j.foodchem.2015.02.030

[99] Giuberti G, Gallo A, Cerioli C, Fortunati P, Masoero F. Cooking quality and starch digestibility of gluten free pasta using new bean flour. Food Chemistry. 2015;**175**:43-49. DOI: 10.1016/j.foodchem.2014.11.127

[100] Giuberti G, Rocchetti G, Sigolo S, Fortunati P, Lucini L, Gallo A. Exploitation of alfalfa seed (*Medicago sativa* L.) flour into gluten-free rice cookies: Nutritional, antioxidant and quality characteristics. Food Chemistry. 2018;**239**:679-687. DOI: 10.1016/j.foodchem.2017.07.004

[101] Xie F, Huang Q, Fang F, Chen S, Wang Z, Wang K, et al. Effects of tea polyphenols and gluten addition on in vitro wheat starch digestion properties. International Journal of Biological Macromolecules. 2019;**126**:525-530. DOI: 10.1016/j.ijbiomac.2018.12.224

[102] Yang Y, Wang L, Li Y, Qian HF, Zhang H, Wu GC, et al. Investigation the molecular degradation, starch-lipid complexes formation and pasting properties of wheat starch in instant noodles during deep-frying treatment.

Food Chemistry. 2019;**283**:287-293. DOI: 10.1016/j.foodchem.2019.01.034

[103] Khatun A. The impact of rice protein and lipid on in vitro rice starch digestibility [Doctoral dissertation]. East Lismore: Southern Cross University; 2019

[104] Chen X, Luo J, Fu L, Cai D, Lu X, Liang Z, et al. Structural, physicochemical, and digestibility properties of starch-soybean peptide complex subjected to heat moisture treatment. Food Chemistry. 2019;**297**:124957. DOI: 10.1016/j.foodchem.2019.124957

[105] Sofi SA, Singh J, Chhikara N, Panghal A. Effect of incorporation of germinated flour and protein isolate from chickpea on different quality characteristics of rice-based noodle. Cereal Chemistry. 2020;**97**(1):85-94. DOI: 10.1002/cche.10192

[106] Gelencser T, Gal V, Hodsagi M, Salgo A. Evaluation of quality and digestibility characteristics of resistant starch-enriched pasta. Food and Bioprocess Technology. 2008;**1**(2):171-179. DOI: 10.1007/s11947-007-0040-z

[107] Capriles VD, Arêas JA. Effects of prebiotic inulin-type fructans on structure, quality, sensory acceptance and glycemic response of gluten-free breads. Food and Function. 2013;**4**:104-110. DOI: 10.1039/C2FO10283H

[108] Bustos MC, Perez GT, Leon AE. Sensory and nutritional attributes of fibre-enriched pasta. LWT-Food Science and Technology. 2011;**44**:1429-1434. DOI: 10.1016/j.lwt.2011.02.002

[109] Giuberti G, Fortunati P, Gallo A. Can different types of resistant starch influence the in vitro starch digestion of gluten free breads? Journal of Cereal Science. 2016;**70**:253-255. DOI: 10.1016/j.jcs.2016.07.001

Chapter 8

Resistant Starch: A Promising Functional Food Ingredient

Revati Wanikar and Swati Kotwal

Abstract

Nowadays dietary starches are considered as a tool for maintaining good health. Recently resistant starch has received much attention because of its specific contribution to human health. Resistant starch escapes digestion in the small intestine and fermented in the colon by colonic microorganisms. Resistant starch has wide applications in varieties of food products. In the present study, types of resistant starch, their sources, physiological benefits, have been discussed briefly. This chapter focuses on factors affecting starch digestion, resistant starch content, characterization of resistant starch and various techniques employed to study their structural features.

Keywords: resistant starch, starch digestion, glycemic index, short chain fatty acids, molecular characterization, SEM

1. Introduction

The concept of resistant starch (RS) has raised interest as a source of dietary fiber. A recent recognition of resistant starch as a functional food ingredient finds application in varieties of food products. The term "Resistant Starch" was first coined by Englyst et al. in 1982 [1] and later defined formally by European Flair Concerted Action on Resistant Starch (EURESTA) as "a fraction of starch that resists digestion in the small intestine of healthy individual and passes to the large intestine where it is a substrate for bacterial fermentation" [2].

RS has potential health benefits similar to soluble fiber. The content of resistant starch in foods has considerable importance because it positively influences functioning of digestive tract, gut microflora, glycemic index, maintain blood cholesterol level and assist in the control of diabetes. These qualities of RS are attracting the attention of food industries and to understand its formation and ways to modulate its content according to the need of the human ailment [3].

History of starch and its usage by man has been extensively studied and are well documented over the years. Starch is the most significant form of carbohydrate in terms of its universality as an energy source in human diet and its applicability in varieties of food products. The understanding that starch is not completely digested and the finding that some starches are poorly digested has led to improved interest for nutritionist. Starch digestion, its impact on glucose release and its relevance to diabetes, obesity and other metabolic disorders resulted in renewed interest in intake of starchy foods. Starchy foods which release glucose slowly and over a longer period

IntechOpen

of time after digestion are of great interest. Controlling glucose release from starchy foods has become challenge for food developers in the context of worldwide health concern. It is possible to modify the structure of starch for desired functional properties by applying various food processing [4–9].

Starch is utilized in several industrial applications due to its ability to impart broad range of functional properties to food and non food products. The new insights have increased the interest in identifying new sources of starches with distinct functional properties and their potential for processing at large scale [10, 11].

Starch is the only natural polysaccharide digested by enzymes of human gastrointestinal tract. Starch digestion starts in mouth where α-amylase in saliva breaks down starch into oligosaccharides and maltose. The bolus is then transported to the stomach where the enzyme activity is inhibited due to low pH and therefore starch does not break down until reaches to the small intestine. By the action of pancreatic α-amylase in the small intestine starch is broken down to glucose and maltose however all the starch is not hydrolysed and absorbed. Fraction of starch which escapes digestion is passed into the large intestine and fermented by intestinal microflora. Hydrolysis of starch by enzymatic digestion may be affected by digestion conditions, granule size, amylose/amylopectin ratio and processing method of starch [12]. Starch is normally processed or cooked before being consumed by humans; hence extent of disruption of starch structure determines its susceptibility to enzymatic digestion [13].

According to in vitro digestion, starch is classified into three categories

1. **Rapidly digestible starch (RDS):** RDS is the fraction of starch that causes rapid increase in blood glucose level after ingestion. Chemically it is measured as a starch digested to glucose after 20 min of α-amylase incubation. RDS is mainly the amorphous fractions of starch and may occur in high amount in freshly cooked foods [3, 14].

2. **Slowly digestible starch (SDS):** SDS is the starch fraction that is digested slowly in the small intestine. SDS cannot be disrupted by salivary α-amylase; it is hydrolysed by pancreatic α-amylase and broken down into linear oligomers and limit dextrins. SDS measured chemically as a starch digested in 100 min of enzyme incubation. Due to slow release of glucose SDS has potential health benefits. SDS reduces risk of chronic diseases related to diet such as diabetes, obesity and other metabolic syndrome [15]. SDS occurs in raw starches with crystalline pattern of A and C type and in the retrograded starch [16].

3. **Resistant starch (RS):** RS is defined as the fraction of starch that escapes digestion in small intestine and fermented in the colon. Chemically, RS can also be defined as fraction of starch not digested after 120 min of incubation with enzymes [1, 14, 16].

2. Types of resistant starch

Depending on its resistance to digestion, RS is classified as RS1, RS2, RS3 and RS4 (**Table 1**).

RS1 is a physically protected starch surrounded by cell wall and other food matrix which hinders the digestibility of starch. RS1 is found in whole or partially milled seeds, cereal grains or legumes. Human gastrointestinal tract lacks the enzymes need to degrade

Type	Description	Food sources	Resistance reduced by
RS I	Physically protected	Whole or partially milled seeds, legumes, pasta	Milling and chewing
RS II	Non gelatinized granules with B-type crystallinity and are hydrolyzed slowly by α-amylase	Raw potatoes, green banana, some legumes, high amylose starch	Food processing and cooking
RS III	Retrograded starch	Cooked and cooled cereal products with prolong and/ or repeated moist heat treatment	Processing conditions
RS IV	Chemically modified starches due to cross-bonding with chemical reagents.	Some fiber drink, foods in which modified starches has been used (e.g. certain breads and cakes)	Less susceptible to *in vitro* digestibility

Source: Nugent [17].

Table 1.
Types of resistant starch.

cellulose, hemicelluloses, lignin and other plant cell wall constituents and therefore this form of physically protected starch passes to the small intestine in intact form [18].

RS2 is a starch in a certain granular form and they are protected from digestive enzymes due to their crystalline structure. Such type of starch is mostly present in uncooked potatoes and bananas. Raw potato starch has large granule size and hence limited access to the enzymatic attack [18, 19]. The extent of starch hydrolysis is determined by the structure and size of the starch granule surface. However, No relationship has been reported between the extent of starch hydrolysis and degree of enzyme adsorption on the surface of the starch granule. Potato starches have B type crystalline pattern whereas cereal starches are characterized by A type with higher degree of crystallinity and therefore susceptible to enzymatic attack compared to potato starch. Waxy maize starch which contains 100% amylopectin with 40% crystallinity is more susceptible to digestion than high amylose maize starch with 15% crystallinity [20]. Crystallinity plays an important role in the architecture of the granules in terms of its susceptibility to enzymatic hydrolysis.

RS3 is a retrograded starch. Retrogradation occurs when starchy foods are gelatinized and cooled. Gelatinization is a process in which starch is heated in presence of water which resulted in swelling of the granule, leaching of amylose and loss of crystalline structure. Gelatinization is a complex process which starts at low temperature by swelling and continues until the granules are disrupted completely. As the temperature increases the interaction between the polymers decrease and starch granule breaks down. These structural changes take place during heating of starch in the presence of water. Extent of starch gelatinization depends on many factors such as botanical source of starch, heating rate, water content, amylose-amylopectin ratio, and processes applied to starch before gelatinization [21, 22]. Retrogradation is a process in which gelatinized starch upon cooling tends to reassociate to form more ordered structure. This re-annealing of amylose and amylopectin branches occur when gelatinized starch is stored at lower temperatures for longer period of time and thus protects from enzymatic attack [23]. Retrogradation is a property of starch, which is of particular interest in terms of nutritional significance and digestibility. Starch retrogradation was initially thought to be undesirable because of its staling effect on bread and other starchy foods, affecting shelf life and consumer acceptance. However, intensive research on retrogradation of starch over the years have shown that it is desirable in some applications such as preparation of breakfast cereals, parboiled rice, mashed potatoes, chinese rice, because of the changes in structural, sensory and mechanical properties [23]. The most important and significant property of retrograded starch is its slow releaseof glucose into the bloodstream [22, 24]. Retrogradation of starch is associated with series of physical changessuch as increase in viscosity, gel formation, increased degree of crystallinity with the formation of B type crystalline pattern [22]. It is an ongoing process in unstable gelatinized starch, due to rapid recrystallization in amylose polymers followed by slow recrystallization of amylopectin molecules [25].

RS4 is a chemically modified starch formed by cross linking or by adding chemical derivatives.

Recently, two components have been proposed as RS5. The first component is amylose-lipid-complex and second component is resistant maltodextrins [17, 26]. RS occurs naturally in all starchy foods and can be developed in others by combination of several processing conditions.

3. Sources of resistant starch

High amount of resistant starch is found in raw potato and unripe banana. Several studies conferred the beneficial effects of unripe banana on human health which is associated with its high RS content. Raw potato starch has the highest RS content (75%). Whole grains are rich sources of dietary fiber and resistant starch. **Table 2** provides RS content of some basic foods [27].

Source	Total starch	Total dietary fiber	Resistant starch
Legumes	42.6	36.8	24.6
Red kidney beans	53.3	33.3	25.4
Lentils	53.9	32.8	17.7
Black-eyed peas			
Cereal grains	55.5	17.0	18.2
Barley	77.9	19.6	25.2
Corn	50.0	17.0	13.6
Wheat	95.1	1.5	14.1
White rice			
Cereal products	67.4	n/a	1.4
Crisp bread	46.7	n/a	1.9
White bread	67.0	n/a	1.2
Puffed wheat cereal			

Source: J. Lunn et al. [27].

Table 2.
RS content of some basic foods (g/100 g).

4. Nutritional and health impact of resistant starch

4.1 RS as a prebiotic agent

'Prebiotics' are food ingredients that help support growth of probiotic bacteria. Prebiotics are considered as nondigestible carbohydrates such as resistant starch which ferment in the colon by gut microflora. Essentially they stimulate activity of good bacteria such as *Lactobacilli, Bifidobacteria* and *Staphylococci* and confer benefits upon host health [22].

4.2 Prevention of colon cancer

Resistant starch escapes digestion in the small intestine and is fermented in the large intestine resulting in the production of short chain fatty acids (SCFA), some gases like methane, hydrogen and carbon dioxide and organic acid (e.g. lactic acid) [28]. SCFAinclude acetate, propionate and butyrate. A number of studies have indicated the benefits of resistant starch as it produces SCFA, as compared to dietary fiber, especially butyrate production is more. Butyrate is the main energy substrate for colonocytes and several in vitro studies have shown that butyrate inhibits malignant transformation of cells by arresting one of the phases of cell cycle (G1) [23, 25]. More butyrate production is associated with lower incidence of colon cancer [29, 30]. **Table 3** presents data on SCFA produced by the fermentation of some foods in the large intestine.

4.3 Hypoglycaemic effects

Foods containing high resistant starch reduce the rate of digestion. Slow rate of digestion has implications for the use of RS in controlled glucose release applications. Starch digestion and concurrent changes in blood glucose levels are largely dependent on its rate of hydrolysis by α-amylase and extent of digestion. From the health point of view, the starches that are less susceptible to α-amylase attack score high as they bring about less change in post prandial glucose level and more starch enters the colon undigested.RS consumption is associated with reduced post prandial glycemic and insulinemic response. Therefore RS can help in the treatment of diabetes, obesity and in weight management [30].

4.4 Hypocholesterolemic effects

Based on the studies in rats, RS is shown to affect lipid metabolism where reductions in measures of lipid metabolism is observed (total lipids, total cholesterol, LDL, HDL, VLDL) [30].

Substrate	SCFA (%)			
	Acetate	Propionate	Butyrate	References
Resistant starch	41	21	38	35
Oat bran	57	21	23	35
Wheat bran	57	15	19	35
Cellulose	61	20	19	35
Guar gum	59	26	11	35
Pectin	75	14	9	35

Source: A. Sharma et al. [31].

Table 3.
Percentage of total SCFA produced by various substrates.

4.5 Inhibition of fat accumulation

Various studies examined that high RS meals may increase the use of fat stores as a result of reduction in insulin secretions. High RS meals imparted less satiety than low RS meals whereas in another study on human volunteers, high RS meal caused greater satiety [32]. Keenan et.al reported in their study that incorporating RS in diet may increase the gut hormones that are effective in reducing energy intake. This may be an effective approach for the treatment of obesity [33].

4.6 RS as a functional ingredient

The functional properties of resistant starch such as swelling, viscosity, gel consistency, water holding capacity make it useful in variety of food applications. Low water holding capacity of RS makes it a functional ingredient which provides good handling in processing, crispness, expansion and improved texture of food products [14, 19]. Hi-maize is the first commercial RS introduced in the market in 1993 in Australia. The other sources of commercial RS 3 are CrystaLean, Novelose and Actistar which are highly retrograded starches. Fibersym is a chemically modified RS 4 product [14]. RS may find applications in varieties of food products such as bakery products [34, 35], pastas and puddings [14, 36, 37], yoghurt, cheese, icecreams [19, 38, 39]. RS incorporated biscuits has been investigated and reported that incorporating RS in foods have potential to develop fiber rich products without changing their general properties. RS can also be used as thickening agent and substituted fat in imitation cheese and many other products where insoluble fiber is desirable conferring the benefits of RS as a functional fiber. Bread and pasta are the most widely consumed starch based products. RS as a food ingredient is increasingly important as resistant starch has low calorific value (8 kJ/g) compared to fully digestible starch (15 kJ/g) [18].

5. Factors affecting starch digestibility and resistant starch content

The structural changes of starch during processing are the major determinants of starch functional properties for food processing, during digestion and in industrial applications [25, 40]. Wide range of techniques has been used for processing the food materials which involve chemical and hydrothermal treatments. The processing methods are reported to influence the nutritional characteristics of foods. Roasting and cooking without pressure are some of the major processes used in household whereas domestic storage is also a widely used method now-a-days. Processing methods are the major determinants of starch digestibility and amount of starches reaching the colon [41]. Gelatinization and retrogradation are important properties of starch that determine its functionality, quality, acceptability and nutritional value [22]. Several inherent properties of starch influence the formation of RS and starch digestibility are discussed below.

5.1 Granule morphology

Size and shape of starch granule is influenced by botanical origin. Several studies have indicated negative relationship between large granule size of wheat, barley, and potato and starch digestibility. The rate of starch hydrolysis is increased by decreasing the size of the granule. This was observed among starches with different botanical origin [15]. Smaller granules have the higher susceptibility to enzyme binding [42].

5.2 Surface of granules

Starch hydrolysis is also dependent on the shape of the granules which varies from spherical to polyhedral. The molecular association of starch granules may reduce the binding of amylase to granule surface [43].

The surface characteristics such as pin holes, equatorial grooves, indentations and small nodules have an impact on starch digestibility [12] smooth surface of potato and high amylose starches with few pits and pores explain the starch resistance to amylases [42, 43].

5.3 Molecular structure and crystallinity

The different crystalline patterns of starches such as A, B and C differ in their packing of double helical structures of amylopectin molecules thereby influencing their hydrolysis [44] It is reported that B type crystalline starches are more resistant to amylolytic attack than A type. Amylase attack also depends on linear chain length of amylopectin molecule which forms the helices. The longer chains are more resistant to enzymatic hydrolysis due to more stable helices [12, 15]. The hydrolysis starts earlier in the amorphous region of C type crystalline starches. Additionally, the crystalline distribution in granules has an impact on digestibility. Higher resistance was observed in starches with higher amount of double helices. This may be attributed to the resistance of high amylose native starches, which are less crystalline than native starches with high crystallinity.

5.4 Amylose amylopectin ratio

There is a positive correlation between amylose content and resistant starch formation. The linear amylose chains are bound to each other by hydrogen bonds which make them less accessible to hydrolysis [12]. The high proportion of amylopectin molecule in starch granule makes the larger surface area and therefore a molecule becomes more accessible to amylolytic attack. Starch gelatinization is difficult in high amylose starches and is more susceptible to retrogradation [36]. The *in vitro* and *in vivo* starch digestibility of high amylose starches were reported to be lower than normal starches [45].

5.5 Interaction of starch with other components

Food matrixes such as proteins and lipids play significant role during processing and affect the starch digestibility.

5.5.1 Lipids

Lipids are associated with starch granules. The free fatty acids and phospholipids are complexed with amylose and make the starch resistant to digestion. The lipids are usually present on the surface of the granules and reduce the binding of enzymes. The enzymatic digestibility is also reduced by addition of lauric, palmitic and oleic acid [43, 46].

5.5.2 Proteins

The surface proteins influences enzyme binding and limit the rate of hydrolysis. The starch from pulses is hard to digest due to interaction with proteins and presence of protective network around the granule [12].

5.5.3 Dietary fiber

Gaur and xanthan gums are some of the dietary fibers which affect the digestibility due to their high viscosity which slows down the movement and absorption of digestion products in the small intestine [14].

5.5.4 Ions

Phosphorous as phosphate monoesters and phospholipids significantly affect the starch properties. The tendency of phospholipids to form complexes with amylose and amylopectin makes the starch less susceptible to enzyme hydrolysis. Calcium and potassium ions reported to decrease RS yield [47].

6. Techniques used to study morphological, molecular and thermal characteristics of resistant starch

Understanding the molecular characteristics of starches to study the functional behavior and their suitability and applicability in various food industries is of great importance.

Thermal property is an important functional property of starch that varies with respect to the macromolecular composition (amylose and amylopectin ratio), double helical structure of amylopectin (chain length, branching, and degree of polymerization) and granule architecture (amorphous to crystalline ratio), granule morphology and size distribution. Differential scanning calorimetry (DSC) is the widely used technique to study thermal behavior of starches as well as other polymers. DSC can characterize modifications in starches, high amylose starches and waxy maize starches as well [48].

Spectroscopic techniques can provide appropriate information about the native as well as modified starches and their structural features. It also provides information of structural changes during gelatinisation and retrogradation.

The infrared (IR) spectroscopy can detect the molecular bond vibrations (especially C—O and C—C bonds) which yield both qualitative and quantitative information, such as that on the amorphous and crystalline regions of the starch granule [49]. Using FT-IR technique it was also observed that the high amylose maize and potato starches (RS2) exhibited greater level of ordered structure in the external region than wheat, maize or waxy maize starches. Due to retrogradation during storage, conformational changes in starches can be monitored and the intensity changes of conformational-sensitive bands in the 1300–800 cm^{-1} region could be observed [23].

Scanning electron microscopy technique is generally used to provide topographic features of RS. Differences in granule morphologyof starches can also be detected using SEM [48, 50]. SEM images of native versus resistant starches formed from different processing techniques are shown in **Figure 1**. Pinholes on the surface of the starch granules were observed in native starches isolated from millets. RS from cooked samples showed irregular and uneven surface zone. RS from retrograded starches showed fibrous, compact and less smooth structures [51].

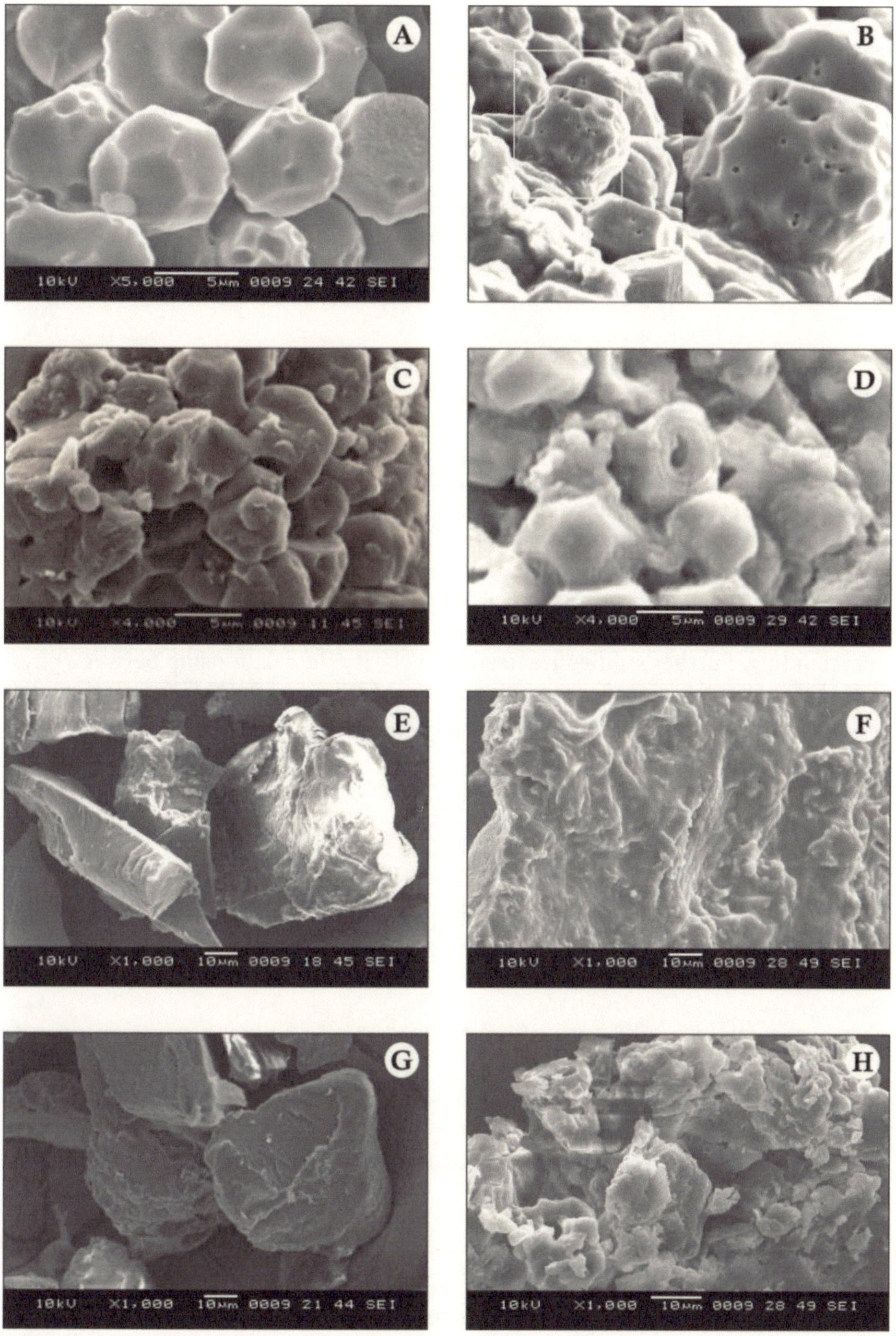

Figure 1.
SEM image of native versus resistant starches. Note: (A) TS native, (B) RS native, (C) TS roasted, (D) RS roasted, (E) TS cooked, (F) RS cooked, (G) TS cooked and stored at −20°C for 30 days, (H) RS cooked and stored at −20°C for 30 days. TS: total starch; and RS: resistant starch. Source: R. Wanikar [51].

X-ray diffraction can be applied to investigate different pattern of crystalline Structure and crystallinity of starch obtained from various botanical sources. XRD generally detects the regularly repeating ordering of helices and thereby reflecting the three-dimensional order of crystalline structure of starch [52].

There is an increased awareness in consumers for health and diet which has led enormous research on resistant starch, its content in foods and structural characterization. When combined the information generated from the above techniques can provide comprehensive analysis of structural characteristics of resistant starch, as well as changes occur during the formation of RS when compared with the structure of their native starches.

7. Conclusion

Resistant starch is not accessible to digestive enzymes. This undigested starch fraction is of particular significance to human health as it lowers the calorific value of food and therefore provides a means to use as a potential food ingredient. The content of resistant starch can be increased by various food processing. Consumer's awareness about health and food is one of the reasons for increased popularity of extensive research on resistant starch and their health impact. Structural characterization of RS by using different techniques and their relationship needs a deeper understanding. Further studies are needed to clarify the relationship between physiological effects and molecular characterization of RS. In vitro RS fermentation and colon cancer incidence is an important aspect for further study.

Author details

Revati Wanikar[1*] and Swati Kotwal[2]

1 Department of Biotechnology, Dr. D.Y.Patil ACS College, Savitribai Phule Pune University, Pimpri, Pune, India

2 Department of Biochemistry, RTM Nagpur University, Nagpur, India

*Address all correspondence to: wanikarrevati@gmail.com

IntechOpen

© 2021 The Author(s). Licensee IntechOpen. This chapter is distributed under the terms of the Creative Commons Attribution License (http://creativecommons.org/licenses/by/3.0), which permits unrestricted use, distribution, and reproduction in any medium, provided the original work is properly cited. (cc) BY

References

[1] Englyst HN, Kingman S, Cummings J. Classification and measurement of nutritionally important starch fractions. European Journal of Clinical Nutrition. 1992;**46**:S33-S50

[2] Englyst HN, Cummings JH. Resistant starch, a new food component: classification of starch for nutritional purposes. In: Morton ID, editor. Cereals in a European Context. First European Conference on Food Science and Technology. Chichester: EllisHorwood; 1987. pp. 221-233

[3] Annison G, Topping DL. Nutritional role of resistant starch: chemical structure vs physiological function. Annual Reviews in Nutrition. 1994;**14**:297-320

[4] Eliasson A-C. Starch in Food, Structure, Function and Application. USA: Woodhead Publishing Limited, CRC Press; 2004. Available from: www.woodhead-publishing.com

[5] Vinoy S et al. Slow release carbohydrates: Growing evidence on metabolic responces and public health interest. Food Nutrition Research. 2016;**60**:31662

[6] Taylor JRN et al. Developments in modulating glycemic response in starchy cereal foods. Starch/Staerke. 2015; **67**(1-2):79-89

[7] Alay SCA et al. Physiological properties, modifications and applications of starches from different botanical sources. Food Science and Technology. 2015;**35**(2):215-236

[8] Tharanathan RN. Starch—Value addition by modification. Critical Reviews in Food Science and Nutrition. 2005;**45**:371-384

[9] Park SH et al. Properties and applications of starch modifying enzymes for use in the banking industry. Food Science and Biotechnology. 2018;**27**(2):299-312

[10] Ádina L, Santana M, Meireles AA. New starches are the trend for industry applications: A review. Food and Public Health. 2014;**4**(5):229-241

[11] SubaricaAckar D, Jurislav Babic BM. Medicinski Glasnik. Starch for Health. 2011;**9**:17-22

[12] Singh J et al. Starch digestion in food matrix. Trends in Food Science and Technology. 2010;**21**:168-180

[13] Wang TL, Bogracheva TY, Hedley CL. Starch: as simple as A, B, C? Journal of Experimental Botany. 1998;**49**(320):481-502

[14] Sajilata MG, Singhal RS, Kulkarni PR. Resistant starch—A review. Comprehensive Reviews in Food Science and Food Safety. 2006;**5**(1):1-17

[15] Lehmann U, Robin F. slowly digestible starch—Its structure and health implications: a review. Trends in Food Science & Technology. 2007;**18**(7): 346-355

[16] Dona AC, Pages G, Gilbert RG, Kuchel PW, et al. Digestion of starch: In vivo and in vitro kinetic models used to characterise oligosaccharide or glucose release—Review. Carbohydrate Polymers. 2010;**80**:599-617

[17] Nugent AP. Health properties of resistant starch. Nutrition Bulletin. 2005;**30**(1):27-54

[18] Leszczyñski W. Resistant starch–classification, structure, production.

Polish Journal of Food and Nutrition Sciences. 2004;**13**(54):37-50

[19] Fuentes-Zaragoza E, Sánchez-Zapata E, Sendra E, Sayas E, Navarro C, Fernández-López J, et al. Resistant starch as prebiotic: A review. Starch-Stärke. 2011;**63**(7):406-415

[20] Brown I. Complex carbohydrates and resistant starch. Nutrition Reviews. 1996;**1996**(54):S115-S119

[21] Ring et al. Some studies on starch gelation. Starch-Strake. 1985;**37**: 8083

[22] Wang S, Copeland L. Molecular disassembly of starch granules. Food & Function. 2013;**4**:1564

[23] Karim AA, Norziah MH, Seow CC. Methods for the study of starch retrogradation. Food Chemistry. 2000;**71**:9-36

[24] Haralampu S. Resistant starch-a review of the physical properties and biological impact of RS 3. Carbohydrate Polymers. 2000;**41**(3):285-292

[25] Wang S et al. Starch retrogradation: A comprehensive review. Comprehensive Reviews in Food Science and Food Safety. 2015;**14**:568-585

[26] Lockyer S, Nugent AP. Health effects of resistant starch. Nutrition Bulletin. 2017;**42**(1):10-41

[27] Lunn J, Buttriss JL. Carbohydrates and dietary fibre. Nutrition Bulletin. 2007;**32**:21-64

[28] Cummings JH, Beatty ER, Kingman SM, Bingham SA, Englyst HN, et al. Digestion and physiological properties of resistant starch in human large bowel. British Journal of Nutrition. 1996;**75**(5):733-747

[29] Cassidy A, Bingham SA, Cummings JH. Starch intake and colorectal cancer risk: an international comparison. British Journal of Cancer. 1994;**69**:937-942

[30] Birt DF, Boylston T, Hendrich S. Resistant starch: promise for improving human health. Advances in Nutrition. 2013;**6**:587-601

[31] Sharma A, Yadav BS, Ritika BY. Resistant starch: Physiological roles and food applications. Food Reviews International. 2008;**24**:193-234

[32] Anderson GH, Catherine NL, Woodened DM. Inverse association between the effect of carbohydrates on blood glucose and subsequent short term food intake in young men. American Journal of Clinical Nutrition. 2002;**76**: 1023-1030

[33] Keenan MJ, Zhou J. Effect of resistant starch, a non digestible fermentable fiber, on reducing body fat. Obesity. 2006;**14**:1523-1534

[34] Korus J, Witczak M, Ziobro R, Juszczak L. The impact of resistant starch on characteristics of gluten-free dough and bread. Food Hydrocolloids. 2009;**23**(3):988-995

[35] Sanz-Penella JM, Wronkowska M, Smietana MS, Collar C, Haros M. Impact of the addition of resistant starch from modified pea starch on dough and bread performance. European Food Research and Technology. 2010;**231**:499-508

[36] Gelencsér T, Juhász R, Hódsági M, Gergely S, Salgó A. Comparative study of native and resistant starches. Acta Alimentaria. 2008;**37**(2):255-270

[37] Ares G, Baixauli R, Sanz T, Varela P, Salvador A. New functional fibre in milk puddings: Effect on sensory properties

and consumers acceptability. LWT - Food Science and Technology. 2009;**42**:710-716

[38] Homayouni A, Azizi A, Ehsani MR, Yarmand MS, Razavi SH. Effect of microencapsulation and resistant starch on the probiotic survival and sensory properties of synbiotic ice cream. Food Chemistry. 2008;**111**:50-55

[39] Arimi JM, Duggan E, O'Riordan ED, O'Sullivan M, Lyng JG. Microwave expansion of imitation cheese containing resistant starch. Journal of Food Engineering. 2008;**88**:254-262

[40] Englyst KN, Hudson GJ, Englyst HN. Starch analysis in food. In: Encyclopedia of Analytical Chemistry. Englyst Carbohydrate services, Eastleigh, UK: Wiley online library; 2000. pp. 1-19

[41] Muir JG, Birkett A, Brown I, Jones G, O'Dea K. Food processing and maize variety affects amounts of starch escaping digestion in the small intestine. The American Journal of Clinical Nutrition. 1995;**61**:82-89

[42] Tester RF, Qi X, Karkalas J. Hydrolysis of native starches with amylases. Animal Feed Science and Technology. 2006;**130**:39-54

[43] Tester RF, Karkalas J, Qi X. Starch—Composition, fine structure and architecture—Review. Journal of Cereal Science. 2004;**39**(2):151-165

[44] Li L, Jiang H, Campbell M, Blanco M, Jane J. Characterization of maize amylose-extender (ae) mutant starches, Part I: Relationship between resistant starch contents and molecular structures. Carbohydrate Polymers. 2008;**74**(3):396-404

[45] Topping DL, Clifton PM. Short-chain fatty acids and human colonic function: Roles of resistant starch and non-starch polysaccharides. Physiological Reviews. 2001;**81**:1031-1064

[46] Svihus B, Uhlen AK, Harstad OM. Effect of starch granule structure, associated components and processing on nutritive value of cereal starch: A review. Animal Feed Science and Technology. 2005;**122**:303-320

[47] Escarpa A, Gonzalez MC, Morales MD, Saura-Calixto F, et al. An approach to the influence of nutrients and other food constituents on resistant starch formation. Food Chemistry. 1997;**60**(4):527-532

[48] Singh J, Singh N. Studies on the morphological, thermal and rheological properties of starch separated from some Indian potato cultivars. Food Chemistry. 2001;**75**:67-77

[49] Millan-Testa CE, Mendez-Montealvo MG, Ottenhof MA, Farhat IA, Bello-Pérez LA. Determination of the molecular and structural charactersitics of okenia, mango, and banana starches. Journal of Agricultural and Food Chemistry. 2005;**53**:495-501

[50] Lindeboom N, Chang PR, Tyler RT. Analytical, biochemical, physicochemical aspects of starch granule size, with emphasis on small granule starches: A review. Starch/Stärke. 2004;**56**(3-4):89-99

[51] Wanikar R, Upadhyay A, Kotwal S. Effect of processing on granular and thermal properties of starch and resistant starch from kodo and kutki. International Journal of Food Sciences and Nutrition. 2018;**7**:83-90

[52] Gidley MJ, Bociek SM. Molecular organization in starches: a carbon 13 CP/MASNMR study. Journal of the American Chemical Society. 1985;**107**(24): 7040-7044

Chapter 9

Starch Biodegradable Films Produced by Electrospraying

Verónica Cuellar Sánchez, Marcela González Vázquez, Alitzel B. García-Hernández, Fátima S. Serrano-Villa, Ma. de la Paz Salgado Cruz, Arturo García Bórquez, Eduardo Morales-Sánchez, Reynold R. Farrera-Rebollo and Georgina Calderón-Domínguez

Abstract

The use of particles obtained from biopolymers is of interest in fields such as bioengineering and nanotechnology, with applications in drug encapsulation, tissue engineering, and edible biofilms. A method used to obtain these particles is electrohydrodynamic atomization (EHDA), which can generate different structures depending on the process conditions and raw materials used, opening a wide range of research in the biopolymers field, where starch is considered an excellent material to produce edible and biodegradable films. This chapter is a compilation and analysis of the newest studies of this technique, using starch with or without modifications to prepare films or membranes and their potential applications. A systematic literature review, focused on starch, and EHDA was carried out, finding 158 articles that match these criteria. From these results, a search inside them, using the words edible and biodegradable was conducted, showing 93 articles with these key words. The information was analyzed observing the preference to use corn, potato, rice, and cassava starches, obtaining mainly scaffolds and fibers and, in much less proportion, films or capsules. This review shows a window of opportunity for the study of starchy materials by EHDA to produce films, coatings, and capsules at micro or nano levels.

Keywords: starch, electrospraying, electrospinning, edible films, biodegradable films

1. Introduction

Over the last decade, due to its multidisciplinary nature, the field of nanotechnology has seen a sharp increase in its applications in several areas, mainly on the "bottom-up" and "top-down" approaches. These terms refer to the synthesis processes used to produce new or modified materials, scaling up atom by atom to form a larger product structure or breaking apart larger particles into micro/nanomaterials, respectively [1]. One of the most reported of these methods, used in the food industry, tissue, and environmental engineering, is the electrohydrodynamic atomization technique (EHDA) [2–4], a "bottom-up" nanotechnology approach, which has been employed for the production of membranes, particles, encapsulation, and edible or biodegradable films.

When EHDA began to be used for the design of micro fibers, nano fibers, and membranes, many of the products were developed with synthetic polymers, which continue giving very good results to this day. However, as the need for greener technologies and more ecofriendly products increased, the use of biopolymers also rose. Thus, carbohydrates such as cellulose, pectin, chitosan, alginate, and starch, single or blended with other bio or synthetic polymers, have increasingly been proposed for the production of films, membranes, fibers, and encapsulates. Among these carbohydrates, starch represents a very good option, as it can be found in large quantities in nature, besides being an inexpensive biopolymer, normally found in leaves, stems, seeds, roots, and tubers or other sources such as algae and bacteria.

In this chapter, we present the basis of EHDA technology and summarize some of the data reported in the most recent studies for the production of fibers, films, and membranes using starch as raw material and analyzing the modifications required to be able to generate these starchy products.

2. Electrohydrodynamic atomization (EHDA)

EHD procesess encompasses two methods called electrohydrodynamic spinning and electrohydrodynamic atomization, better known as electrospinning and electrospraying, respectively [5]. Electrospinning allows for the production of membranes from electrospun fibers, and electrospraying allows for the synthesis of materials such as core/shell, micro/nanoparticles, encapsulates, and films from fine droplets.

A typical EHDA device (**Figure 1**) consists of four parts: (1) a high-voltage power supply (typically ranging from 1 to 30 kV), (2) a syringe pump, (3) a capillary containing the conductive polymer solution (commonly a syringe with a stainless-steel

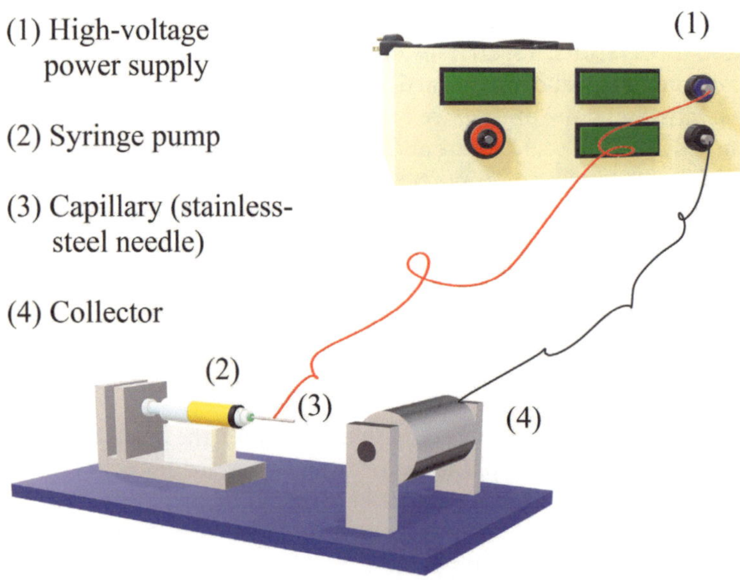

Figure 1.
Electrohydrodynamic equipment. Basic components.

needle), and (4) a collector (stainless-steel rotatory drum or static conductive plate) [2]. These components are present regardless of the method. Moreover, depending on the material to be synthesized, the equipment can be set in two standard configurations (**Figure 2**): horizontal (**Figure 2a**) or vertical (**Figure 2b**) [6], which have been used in the production of films formed by micro/nanoparticles [7] and encapsulates [8] in dry (**Figure 2c**) or wet (**Figure 2d**) configurations.

On the other hand, and in addition to the two standard configurations mentioned above, several modifications have been studied. These modifications have been done according to specific needs; for example, horizontal dry electrospinning (**Figure 2b**) is used to obtain membranes based on hydrolyzed collagen and polyvinyl alcohol with potential use for wound protection [9], and vertical wet spinning (**Figure 2c**) is used to synthesize membranes from polyvinyl alcohol and poly(ethyleneimine), to remove heavy metals from wastewater [10]. In these examples, the collector can be either immersed in a liquid, or dry, (**Figure 2d**).

Characteristics such as product morphology and size are affected by the properties of the solution (viscosity, polymer concentration, molecular weight of polymer, surface tension, conductivity), process variables (applied voltage, working-distance from needle to collector, flow rate), and environmental parameters (temperature, humidity, airflow) [11], resulting in products with different properties and intended uses.

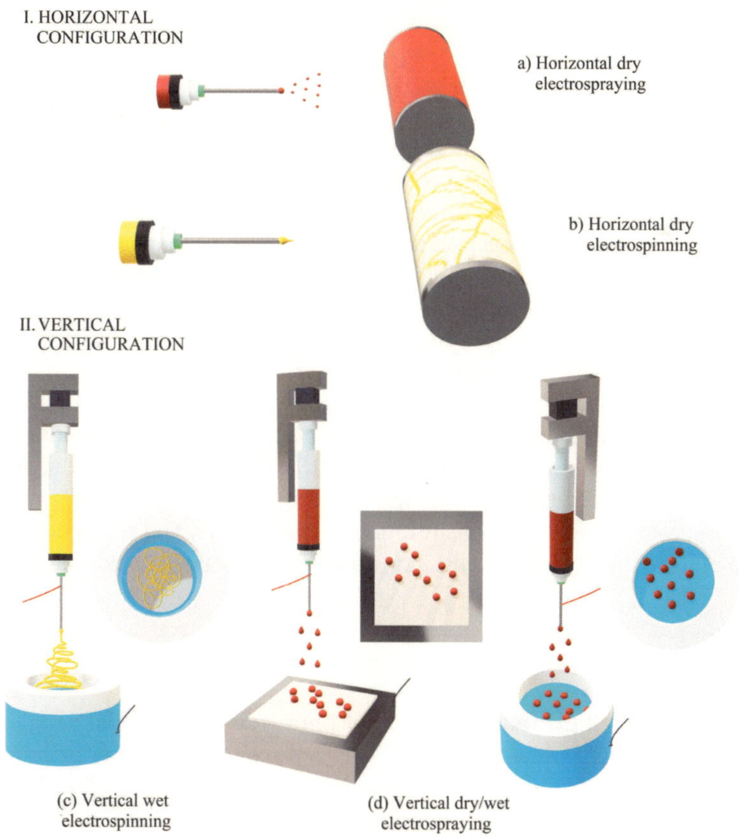

Figure 2.
Electrohydrodynamic configurations: (I) horizontal, (II) vertical. Electrohydrodynamic types: (a) horizontal dry electrospraying, (b) horizontal dry electrospinning, (c) vertical wet electrospinning, (d) vertical dry/wet electrospraying.

But, how are fibers or particles formed? In the case of fibers, when the electrical voltage is applied to the conductive polymer solution in the syringe, electrical charges accumulate on the surface of the liquid and, depending on the surface tension, the polymer solution remains within the capillary, not flowing. As the mutual repulsion of charges produces a force directly opposite to the surface tension and the intensity of the electric field increases, the solution reaches the end of the capillary, acquiring a conical shape, called a "Taylor cone." Consequently, when the electric field reaches a critical value, that is, when the repulsive electrical force exceeds the surface tension force, a jet of the polymer solution is produced at the tip of the cone. As the jet spreads through the air, the solvent in the solution evaporates, forming a polymeric micro or nanofiber. Finally, the fibers are deposited in the collector in the form of a nonwoven micro/nanofiber membrane [6–11].

Regarding the synthesis of films, micro/nanoparticles, and encapsulates, unlike membranes, these materials are formed from solutions of low polymer concentration, which allows the jet's destabilization and the formation of highly identically charged fine droplets that do not agglomerate. In other words, a polymeric solution in the capillary is sprayed from the nozzle into the collector under a high-voltage application due to electrostatic forces. Here, on the flight in time to the collector, the solvent evaporates and particles are produced [11, 12].

The fibers and the particles produced by these methodologies show a high surface area to volume ratio, good mechanical, electrical, and thermal properties, and smooth, homogenous, and variable morphologies, mainly as a result of the process parameters' manipulation, which in turn derives in the shape that the jet takes during the ejection process [13].

As mentioned earlier, the parameters that govern the EHDA process are properties of the solution, process conditions, and environmental parameters, all of which determine the morphology and diameter of the fibers or particles [11, 14].

EHDA products can be synthesized using a wide range of materials, including biopolymers from animals, plants, and algae, such as collagen, chitosan, gelatin, pectin, zein, cellulose, alginate, starch, and others, and synthetic polymers, such as polyethene oxide, polyvinyl alcohol, and polycaprolactone, among others [5, 6, 13]. However, due to their low molecular weight and mechanical properties, natural and synthetic polymers are commonly used in tandem. Furthermore, materials such as carbon-based nanomaterials, ceramics, and metallic nanoparticles have also been applied in combination with chitosan or casein nanofibers, to name a few [15, 16]. In general, since its invention, the application of the EHDA technique increased considerably, due to it being straightforward, inexpensive (low solution consumption), controllable, and reproducible [17], with starch being considered a potential raw material to be used in this technique.

3. Starch

Starch is found in all plants as a product of photosynthesis and is the main storage reserve carbohydrate of plants and the primary source of calories in the human diet. It is also a very important renewable and biodegradable raw material for the industry [18]. The main sources of starch are cereals (corn, wheat, rice, barley) and tubers or roots (potatoes, tapioca, cassava) [19], corn being the most important, followed by potato and cassava.

Starch is a polysaccharide composed of α-glucose polymer molecules: a linear one called amylose and a branched one known as amylopectin. The proportion of these molecules varies depending on the source, with the most common being an amylose content of 13–30%. However, it is possible to find amylopectin-only materials [20, 21], mainly cereals, referred to as waxy cereal varieties (corn, sorghum, rice). These differences in starch composition result in diverse physicochemical properties, affecting properties such as gelatinization temperature, solubility, and final viscosity of starch slurries.

Starch can be extracted by different methods, most of them being classified as dry or wet, and in both cases looking to maintain its functional properties at the highest possible yields and purity [22] and without damaging the crystalline phase or promoting depolymerization [23] of the materials. One of these methods is dry milling, which consists of the grinding of the samples and an air classification [24]. This method simplifies the handling of large amounts of liquid in comparison to wet milling [22] but increases the proportion of damaged starch [25], resulting in a lower quality product [26].

Conversely, wet milling is used to extract starch from flour by producing an aqueous slurry, which is filtrated and washed at least two times [27]; the starch obtained in this process has a higher purity than dry milling [28]. In most wet extraction processes, a reactant, such as sodium bisulfite [29], metabisulfite [30], sodium hydroxide [31], oxalic acid/ammonium oxalate [32], or low concentrations of citric acid [33], is added, mainly to facilitate protein separation. Other techniques, such as sonication [34] or freezing, to assist the extraction process to increase the starch yields have been reported as well.

3.1 Starch sources

Starch is organized into tiny particles called grains or starch granules, and their size and shape are characteristic of each botanical species (**Table 1**). It is known that the granule size is decisive in its processability, which affects the solubility (in a plasticizer medium) and the swelling power, facilitating the release of soluble polymer chains for the formation of a single coherent amorphous phase [47–49].

The size of the starch granule varies from a very small size (4 μm or less), such as that found in amaranth, jicama, or rice, up to 100 μm from potato granules [21]. Most of the materials do not present a unique size and, in some cases, have very different shapes. As an example, in barley starch, there are two populations of granules: small

Type	Amylopectin (%)	Amylose (%)	Gelatinization temperature (°C)	Granule size (microns)	References
Corn	66.19	33.81	70–80	5–25	[21, 35–37]
Corn rich in amylose	20–45	55–80	67–80	5–25	[21]
Potato	79	21	58.5	5–100	[21, 38, 39]
Rice	83	17	68.4–73.95	2–5	[21, 36, 37]
Tapioca	21.4–35.4	24.3	64.1	5–20	[21, 40, 41]
Wax corn	99–100	1–2	64.3–68	5–25	[21, 42]
Wax sorghum	99–100	7.9–12.1	67–74	4–35	[43, 44]
Wheat	76	1.5–39.5	56.1	11–41	[21, 40, 45]
Jicama	73.8	26.4	66.6	3–21	[46]

Table 1.
Some starch characteristics.

2–5 micron-long spheres and large 15–25 micron-long lenticular granules [49]. In the case of rice, corn, and waxy corn starches, they have a polyhedral shape, while the granules of potato starch are ovoid. Cassava follows a similar behavior; starch granules are not uniform, are round with truncated terminals, have a well-defined nucleus, and their size varies between 4 and 35 μm with an average of 20 μm [50, 51]. These differences in size, as well as in amylose and amylopectin content, promote the various functional properties of the starch, such as gelatinization temperatures and thus lead to different industrial applications.

3.2 Modified starches

Starch has many applications in food and nonfood industries based on its physicochemical and functional properties; for example, it is used in the pharmaceutical industry as a raw material for the production of dextrose and serum, as an excipient in the manufacture of tablets and pills, and as capsules [52]. It has been also used as an adhesive, binder, thickener, and co-builder; in gelling, complexing, and flocculating agents; and in the paper and corrugating industry. Another application is in the preparation of edible and biodegradable films, due to barrier characteristics (O_2 and CO_2). However, most of these applications are carried out employing modified starches [21, 52–54].

Starches have functional properties that can be related to their final use and vary depending on the granule secondary and tertiary structures and if the starch has been modified or remains native. These differences influence the gelatinization temperature, type of diffraction patterns, crystallinity degree, solubility, clarity, viscosity, water-retention capacity, and swelling capacity, which help to explain the stability of the biopolymer, and therefore suggest its proper application [55, 56].

Starch can be modified by different procedures, either physical or chemical, reaching different final properties and characteristics. The most common physical modifications include heating starch slurries in boiling water or autoclaving at 121°C, thus promoting gelatinization (low and high temperatures) and as a consequence an increase in its solubilization capacity [56]. Other common physical procedures include ultrasonication [57] and ball milling [58]. Regarding chemical modifications, these procedures change the starch structure, by excising the molecule during a hydrolysis process or by introducing new components as a result of oxidation, esterification, or etherification [53], increasing in most of the cases its solubility and a loss of crystallinity [51, 54].

4. EHDA starch films

Many studies have been carried out regarding electrohydrodynamic atomization, with the first publications about this technique using biopolymers, and specifically starch, coming out in 2003. Many of these documents report on fibers and capsules of different sizes (micro or nano). These were studied alone or as part of scaffolds, membranes, or films—with one or more layers—and built from different polymeric materials besides starch, either of biological or of chemical origin.

Starch is a common material widely distributed in nature, with EHDA products being mainly built from commercial sources, such as corn and maize starch are the ones that have different amylose/amylopectin content [59–67], or others such as potato [66, 68–73], rice [74], and cassava or tapioca starches [75–79].

The use of chemically modified starches, such as cationic starch prepared from hydroxyethylated starch [80], hydroxypropyl starch [81], or octenylsuccinylated starch [82], is also a common practice, while the study of noncommercial biopolymers sources is less frequent [66, 74, 83].

Another normal practice observed for the elaboration of EHDA starch products is combining starch with other polymers, being PVA (polyvinyl alcohol), PCL (polycaprolactone), and PLA (polylactic acid) widely employed [61, 68, 76, 78, 80, 84, 85]. The use of PEO (polyethylene oxide), PMMA (polymethyl methacrylate), and TPU (thermoplastic polyurethane) has also been reported, although in fewer amounts [81, 86].

Starch in its native form is seldom used for the elaboration of EHDA starch products due to its poor solubility and hydrophobicity. This is the reason why it is used in combination with other polymers or modified by physical or chemical procedures.

In this regard, heating by conventional techniques, which render gelatinized starch, is one of the most common procedures. More recently, microwave heating has been reported [74], with both methods increasing the solubility of this polysaccharide. The temperatures reported in these studies use to promote the starch solubilization varied from 70°C up to 140°C, and the heating duration from 10 min to 720 min, with differences seeming to be mostly related to the temperatures used [59–61, 67, 73, 74, 77, 81, 83, 87, 88]. Ultrasonic starch disruption has also been cited [59, 70], along with aqueous DMSO solution to improve starch dissolution [62, 63, 66, 67, 75, 76, 83, 84, 86].

When preparing polysaccharide solutions for electrospinning, the [63] concentrations of native starch [63] are low, ranging from 0.5% [74] to 15%. Higher concentrations of these materials have been reported for commercial soluble (50%) and cassava (66%) starches [71, 87]. In most cases, the solvents added correspond to water [59, 62, 68, 78, 80, 88] or DMSO solutions [63, 70, 75, 76, 83, 87] and in lower amounts to acetic acid, formic acid, ethanol, chloroform, DCM and DMF solutions [71, 79, 84, 85, 89]. **Figure 3** summarized the main steps to prepare starch solutions for electrospraying.

Once the starch solution is obtained, it is fed to the EHDA equipment, and the flow rate, voltage, and distance to collector are set. Most authors reported using

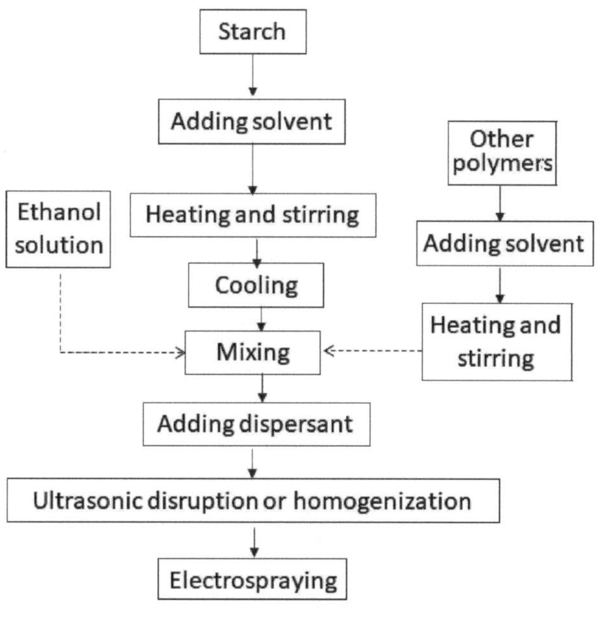

Figure 3.
General method to prepare starch solutions for electrospraying. Dotted lines indicate alternative methodologies.

	EHDA conditions					
Starch source	**DC* (cm)**	**V* (kV)**	**FR* (mL/h)**	**SND* (mm)**	**Product & reference**	
Maize starch (10–20 μm, 27% amylose, 73% amylopectin ratio	2	4.1–7	0.18	0.25	Films	[59]
Cationic starch-PVA	11–14	40–70	NR	NR	Nanofibers	[80]
Starch-PCL	20	9.5	1.0	0.50	Scaffolds	[83]
Corn starch-chitosan-PET	15	20	NR	NR	Fibers	[60]
Cassava Starch-PLA	20	20	0.6	0.55*	Fibers	[75]
Oxidized corn starch-PVA	12	11	NR	0.41	Fibers	[61]
Corn starch of different amylose content-Ming bean starch	5–10	0–15	0.1–0.4	0.60	Fiber entanglements	[83]
Hydroxypropyl starch-PEO	30	11–14	0.02–0.04	0.84	Fiber mats	[81]
Tapioca Starch	15	20	10	0.9	fibers	[77]
Fibersol-guar gum	9-11	10	0.15	NR	Micro/nano capsules	[78]
Potato starch-TPU	24	35	0.75	0.51	Nanofibrous bandages	[84]
Rice starch-Carob flour-PEO	30	12	0.8	NR	Fiber membranes	[74]
Potato starch	20	25	0.6	0.8	Ultrafine fibers	[79]

**Outer diameter.*
DC: distance to collector; V: voltage; FT: flow rate; SND: syringe needle inner diameter. NR: no reported. PVA: polyvinyl alcohol; PCL: polycaprolactone; PET; polyethylene terephthalate; PLA: polylactic acid; PEO: polyethylene oxide; TPU: thermoplastic polyurethane.

Table 2.
Process conditions employed to develop EHDA starch products. Some examples.

voltages between 0 and 20 kV (66%), flow rates smaller than or equal to 1.0 mL/h (81%), and highly variable distances to collector (5–30 cm); in most of these cases, micro and nanofibers or mats were developed, with the exception of two works reporting capsules [86, 87] and two reporting films [59, 67]. However, in some cases, more than one method to prepare mats or films is used, combining, for example, both electrospraying and casting or others [85]. **Table 2** shows some examples of specific process conditions used to obtain the different EHDA starch products.

5. Conclusions

It is of notice that even though starch electrospraying has been studied for many years, most of this research has been focused exclusively into an electrospinning field, with very few works having been published related to the production of edible or biodegradable films, coatings, or microcapsules.

This observation shows a window of opportunity, for the study of new starchy materials and to better understand this technique and its intricacies. Some examples include the effects of different assay parameters, such as syringe inner diameter or the size of starch granules and their relationship to film properties, factors that have not been reported yet. Several studies with other biopolymers [88, 90–92], as well as starch, can serve as a basis for the development of new and improved ecological coating materials.

Acknowledgements

Verónica Cuellar Sánchez, Marcela González Vázquez and Alitzel B. García-Hernández would like to thank CONACyT and BEIFI-IPN for the scholarships provided and the financial support for this work. This research was funded through the projects: 20195500, 20201679, 20201695, 20210624, and 20211381 from the Instituto Politécnico Nacional (IPN, Mexico) and 1668 from CONACyT.

Conflict of interest

The authors declare no conflict of interest.

Author details

Verónica Cuellar Sánchez[1], Marcela González Vázquez[1], Alitzel B. García-Hernández[1], Fátima S. Serrano-Villa[1], Ma. de la Paz Salgado Cruz[1,2], Arturo García Bórquez[3], Eduardo Morales-Sánchez[4], Reynold R. Farrera-Rebollo[1] and Georgina Calderón-Domínguez[1*]

1 Departamento de Ingeniería Bioquímica, Escuela Nacional de Ciencias Biológicas, Instituto Politécnico Nacional, Gustavo A. Madero, México

2 Consejo Nacional de Ciencia y Tecnología (CONACyT), Ciudad de México, México

3 Escuela Superior de Física y Matemáticas, Instituto Politécnico Nacional, Gustavo A. Madero, México

4 Centro de Investigación en Ciencia Aplicada y Tecnología Avanzada, Unidad Querétaro, Instituto Politécnico Nacional, México

*Address all correspondence to: gcalderon@ipn.mx

IntechOpen

© 2021 The Author(s). Licensee IntechOpen. This chapter is distributed under the terms of the Creative Commons Attribution License (http://creativecommons.org/licenses/by/3.0), which permits unrestricted use, distribution, and reproduction in any medium, provided the original work is properly cited.

References

[1] Nanotechnology. 2021. Encyclopedia Britannica [Internet]. Available from: https://www.britannica.com/technology/nanotechnology [Accessed: 8-08-2021]

[2] Anu-Bhushani J, Anandharamakrishnan C. Electrospinning and electrospraying techniques: Potential food based applications. Trends in Food Science and Technology. 2014;**38**:21-33. DOI: 10.1016/j.tifs.2014.03.004

[3] Eltom A, Zhong G, Muhammad A. Scaffold techniques and designs in tissue engineering functions and purposes: A review. Hindawi Advances in Materials Science and Engineering. 2019;**3429527**: 1-13. DOI: 10.1155/2019/3429527

[4] Wang X, Min M, Liu Z, Yang Y, Zhou Z, Zhu M, et al. Poly (ethyleneimine) nanofibrous affinity membrane fabricated via one step wet-electrospinning from poly (vinyl alcohol)-doped poly(ethyleneimine) solution system and its application. Journal of Membrane Science. 2011;**379**:191-199. DOI: 10.1016/j.memsci.2011.05.065

[5] Soares RMD, Siqueira NM, Prabhakaram MP, Ramakrishna S. Electrospinning and electrospray of bio-based and natural polymers for biomaterials development. Materials Science and Engineering: C. 2018;**92**:969-982. DOI: 10.1016/j.msec.2018.08.004

[6] Thenmozhi S, Dharmaraj N, Kadirvelu K, Kim HY. Electrospun nanofibers: New generation materials for advanced applications. Materials Science and Engineering: B. 2017;**217**:36-48. DOI: 10.1016/j.mseb.2017.01.001

[7] Valdespino-León M, Calderón-Domínguez G, Salgado-Cruz MDLP, Rentería-Ortega M, Farrera-Rebollo RR, Morales-Sánchez E, et al. Biodegradable electrosprayed pectin films: An alternative to valorize coffee mucilage. Waste and Biomass Valorization. 2020;**12**:2477-2494. DOI: 10.1007/s12649-020-01194-z

[8] Rentería-Ortega M, Salgado-Cruz MDLP, Morales-Sánchez E, Alamilla-Beltrán L, Farrera-Rebollo RR, Valdespino-León M, et al. Effect of electrohydrodynamic atomization conditions on morphometric characteristics and mechanical resistance of chia mucilage-alginate particles. CyTA - Journal of Food. 2020;**18**:461-471. DOI: 10.1080/19476337.2020.1775706

[9] García-Hernández AB, Morales-Sánchez E, Calderón-Domínguez G, Salgado-Cruz MDLP, Farrera-Rebollo RR, Vega-Cuellar MA, et al. Journal of Applied Polymer Science. 2021:e51197. DOI: 10.1002/app.51197

[10] Sonseca A, Sahay R, Stepien K, Bukala J, Wcislek A, McClain A, et al. Architectured Helically Coiled Scaffolds from Elastomeric Poly(butylene succinate) (PBS) Copolyester via Wet Electrospinning. Materials Science and Engineering: C. 2020;**108**(110505). DOI: 10.1016/j.msec.2019.110505.

[11] Castro-Coelho S, Nogueiro-Estevinho B, Rocha F. Encapsulation in food industry with emerging electrohydrodynamic techniques: Electrospinning and electrospraying - A review. Food Chemistry. 2021;**339**:127850. DOI: 10.1016/j.foodchem.2020.12785

[12] Boda SK, Li X, Xie J. Electrospraying an enabling technology for pharmaceutical and biomedical applications: A review. Journal of

Aerosol Science. 2018;**125**:164-181. DOI: 10.1016/j.jaerosci.2018.04.00

[13] Lim LT, Mendes AC, Chronakis IS. Electrospinning and electrospraying technologies for food applications. In: Advances in Food and Nutrition Research. In: Food Applications of Nanotechnology; 2019. pp. 167-234. DOI: 10.1016/bs.afnr.2019.02.005

[14] Calderón-Arenas JM, Martínez-Rincón HA. Obtención de fibras poliméricas a partir de la técnica de electrospinning para aplicaciones biomédicas [thesis]. Santiago de Cali: Universidad Autónoma de Occidente; 2012. Colombia

[15] Selvaraj S, Thangam R, Fathima NN. Electrospinning of casein nanofibers with silver nanoparticles for potential biomedical applications. International Journal of Biological Macromolecules. 2018;**120**:1674-1681. DOI: 10.1016/j.ijbiomac.2018.09.177

[16] Yan E, Fan S, Li X, Wang C, Sun Z, Ni L, et al. Electrospun polyvinyl alcohol/chitosan composite nanofibers involving Au nanoparticles and their in vitro release properties. Materials Science and Engineering: C. 2013;**33**:461-465. DOI: 10.1016/j.msec.2012.09.014

[17] Jaworek A, Sobczyk AT. Electrospraying route to nanotechnology: An overview. Journal of Electrostatics. 2008;**66**:197-219. DOI: 10.1016/j.elstat.2007.10.001

[18] INECOL. 2021. ¿Qué es el almidón? [Internet]. Available from: https://www.inecol.mx/inecol/index.php/es/ct-menu-item-25/ct-menu-item-27/17-ciencia-hoy/1376-que-es-el-almidon [Accessed: 2021-08-08]

[19] French D. Physical and chemical structure of starch and glycogen. In: Carbohydrates and Their Roles. 1st ed. Westport, Conn: AVI Publishing; 1969

[20] Astiasarán I, Martinez JA. Alimentos: composición y propiedades. Segunda edición. Madrid, Spain: McGraw-Hill Interamericana; 2000. p. 364

[21] Badui DS. Química de los Alimentos. Cuarta edición. Pearson: Addison Wesley; 2006. p. 736

[22] Lee HC, Htoon AK, Paterson JL. Alkaline extraction of starch from Australian lentil cultivars Matilda and Digger optimized for starch yield and starch and protein quality. Food Chemistry. 2007;**102**:551-559. DOI: 10.1016/j.foodchem.2006.03.042

[23] Han X, Hamaker BR. Partial leaching of granule-associated proteins from rice starch during alkaline extraction and subsequent gelatinization. Starch/Stärke. 2002;**54**:454-460. DOI: 10.1002/1521-379X (200210)54:10<454:AID-STAR454>3.0.CO;2-M

[24] Tian S, Kyle WSA, Small DM. Pilot scale isolation of proteins from fields peas (Pisum sativem L.) for use as food ingredients. International Journal of Food Science and Technology. 1999;**34**(33-39). DOI: 10.1046/j.1365-2621.1999.00236.x

[25] Kringel DH, Mello SL, Da Rosa E, Guerra AR. Methods for the extraction of roots, pulses, pseudocereals, and other unconventional starch source: A review. Starch/Stärker. 2020;**72**:11-12. DOI: 10.1002/star.201900234

[26] Steeneken PA, Helmens HJ. Laboratory-scale Dry/wet-milling process for the extraction of starch and gluten from wheat. Starch/Stärker. 2009;**61**:389-397. DOI: 10.1002/star.200800065

[27] Zheng GH, Sosulski FW, Tyler RT. Wet-milling, composition and functional properties of starch and protein isolated from buckwheat groats. Food Research International. 1999;**30**:493-502. DOI: 10.1016/S0963-9969(98)00021-0

[28] Kringel DH, Mello SL, Da Rosa E, Guerra AR. Methods for the extraction of roots, pulses, pseudocereals, and other unconventional starch source: A review. Starch/Stärker. 2020;**72**:11-12. DOI: 10.1002/star.201900234

[29] Lim ST, Lee JH, Shin DH, Lim HS. Comparison of protein extraction solutions for rice starch isolation and effects of residual protein content on starch pasting properties. Starch/Stärker. 1999;**51**:120-125. DOI: 10.1002/(SICI)1521-379X(199904)51:4<120:AID-STAR120>3.0.CO;2-A

[30] Ji Y, Seetharaman K, White PJ. Optimizing a small-scale corn-starch extraction method for use in the laboratory. Cereal Chemistry. 2004;**81**:55-58. DOI: 10.1094/CCHEM.2004.81.1.55

[31] Matsunaga N, Takahashi S, Kainuma K. Rice starch isolation from newly developed rice cultivars by the improved alkali method. Journal of Applied Glycoscience. 2003;**50**:913. DOI: 10.5458/jag.50.9

[32] Daiuto E, Cereba M, Sarmento S, Vilpoux O. Effects of extraction methods on yam (Dioscorea alata) starch characteristics. Starch/Stärker. 2005;**57**:153-160. DOI: 10.1002/star.200400324

[33] Pascoal AM, Di-Medeiros CB, Batista KA, Gonçalves MI, Moraes L, Ferandes KF. Extraction and chemical characterization of starch. Carbohydrate Polymers. 2013;**98**:1304-1310. DOI: 10.1016/j.carbpol.2013.08.009

[34] González LB, Calderón G, Salgado MP, Díaz M, Ramírez M, Chanona JJ, et al. Ultrasound-assisted extraction of starch from frozen jicama (P. erosus) roots: Effect on yield, structural characteristics and thermal properties. CyTA Journal of Food. 2018;**16**:1738-1746. DOI: 10.1080/19476337.2018.1462852

[35] Wang B, Dong Y, Fang Y, Gao W, Kang X, Liu P, et al. Effects of different moisture contents on the structure and properties of corn starch during extrusion. Food Chemistry. 2021;**368**:130804. DOI: 10.1016/j.foodchem.2021.130804

[36] Lutfi Z, Kalim Q, Shahid A, Nawab A. Water chestnut, rice, corn starches and sodium alginate. A comparative study on the physicochemical, thermal and morphological characteristics of starches after dry heating. International Journal of Biological Macromolecules. 2021;**184**:476-482. DOI: 10.1016/j.ijbiomac.2021.06.128

[37] Sun Y, Wang M, Ma S, Wang H. Physicochemical characterization of rice, potato, and pea starches, each with different crystalline pattern, when incorporated with Konjac glucomannan. Food Hydrocolloids. 2020;**101**:105499.7. DOI: 10.1016/j.foodhyd.2019.105499

[38] Datta D, Halder G. Effect of media on degradability, physico-mechanical and optical properties of synthesized polyolefinic and PLA film in comparison with casted potato/corn starch biofilm. Process Safety and Environmental Protection. 2019;**124**:39-62. DOI: 10.1016/j.psep.2019.02.002

[39] Lewandowicz G, Soral-Śmietana M. Starch modification by iterated syneresis. Carbohydrate Polymers. 2004;**56**:403-413. DOI: 10.1016/j.carbpol.2004.03.013

[40] Hsieh CF, Liu W, Whaley JK, Shi YC. Structure, properties, and potential applications of waxy tapioca starches–A review. Trends in Food Science and Technology. 2019;**83**:225-234. DOI: 10.1016/j.tifs.2018.11.022

[41] Cai L, Shi YC. Structure and digestibility of crystalline short-chain amylose from debranched waxy wheat, waxy maize, and waxy potato starches. Carbohydrate Polymers. 2010;**79**:1117-1123. DOI: 10.1016/j.carbpol.2009.10.057

[42] Elhassan MS, Emmambux MN, Hays DB, Peterson GC, Taylor JR. Novel biofortified sorghum lines with combined waxy (high amylopectin) starch and high protein digestibility traits: Effects on endosperm and flour properties. Journal of Cereal Science. 2015;**65**:132-139. DOI: 10.1016/j.jcs.2015.06.017

[43] Ali TM, Hasnain A. Morphological, physicochemical, and pasting properties of modified white sorghum (Sorghum bicolor) starch. International Journal of Food Properties. 2014;**17**:523-535. DOI: 10.1080/10942912.2012.654558

[44] Yuryev VP, Krivandin AV, Kiseleva VI, Wasserman LA, Genkina NK, Fornal B, et al. Structural parameters of amylopectin clusters and semi-crystalline growth rings in wheat starches with different amylose content. Carbohydrate Research. 2004;**339**: 2683-2691. DOI: 10.1016/j.carres.2004.09.005

[45] Ramírez-Miranda M, Ribotta PD, Silva-González AZZ, Salgado-Cruz MDLP, Andraca-Adame JA, Chanona-Pérez JJ, et al. Morphometric and crystallinity changes on jicama starch (Pachyrizus erosus) during gelatinization and their relation with in vitro glycemic index. Starch – Stärke. 2017;**69**:1600281. DOI: 10.1002/star.201600281

[46] Salas C, Medina JA. Caracterización morfológica del gránulo de almidón nativo: Apariencia, forma, tamaño y su distribución. Revista de ingeniería. 2008;**27**:56-62. DOI: 10.16924/revinge.27.6

[47] Jeroen JG, v S and Vliegenthart FGJ. Crystallinity in starch plastics: consequences for material properties. Trends in Biotechnology. 1997;**15**:208-213. DOI: 10.1016/S0167-7799(97)01021-4

[48] Kaur L, Singh N, Singh SN. Some properties of potatoes and their starches II. Morphological, thermal and rheological properties of starches. Food Chemistry. 2002;**79**:183-192. DOI: 10.1016/S0308-8146(02)00130-9

[49] Singh N, Singh J, Kaur L, Singh SN, Singh GB. Morphological, termal and rheologycal properties of starches from different botanical sources. Food Chemistry. 2003;**81**:219-231. DOI: 10.1016/S0308-8146(02)00416-8

[50] Sánchez T, Aristizábal J. Guía técnica para producción y análisis de almidón de yuca. FAO [Internet]2007. p. 129 Available from: http://www.fao.org/3/a1028s/a1028s.pdf. [Accessed: 2021-08-08]

[51] Valdés SE. Hidratos de carbono. En: Badui D. Salvador Química de los Alimentos. Cuarta edición. Pearson: Addison Wesley; 2006. p 29-107.

[52] Röper H. Renewable raw materials in Europe — Industrial utilisation of starch and sugar [1]. Starch-Stärke. 2002;**54**:89-99. DOI: 10.1002/1521-379X(200204)54:3/4<89: AID-STAR89>3.0.CO;2-I

[53] Lee FA. Basic of Food Chemistry. 2nd ed. INC. Westport, Conn: The AVI publishing Company; 1983. p. 546. DOI: 10.1007/978-94-011-7376-6

[54] Hoover R. The Impact of heat-moisture treatment on molecular structures and properties of starches isolated from different botanical sources. Critical Reviews in Food Science and Nutrition. 2010;**50**:835-847. DOI: 10.1080/10408390903001735

[55] Mukerjea R, Slocum G, Robyt JF. Determination of the maximum water solubility of eight native starches and the solubility of their acidic-methanol and -ethanol modified analogues. Carbohydrate Research. 2007;**342**:103-110. DOI: 10.1016/j.carres.2006.10.022

[56] Fernando S, Paranavithana T, Dissanayaka U, Premarathna W, Atambawa A, de Silva N, et al. Effect of starch particle size reduction on the performance of sized warp yarns. Moratuwa Engineering Research Conference (MERCon). 2015:60-63. DOI: 10.1109/mercon.2015.7112321

[57] Dai L, Li C, Zhang J, Cheng F. Preparation and characterization of starch nanocrystals combining ball milling with acid hydrolysis. Carbohydrate Polymers. 2018;**180**:122-127. DOI: 10.1016/j.carbpol.2017.10.015

[58] Pareta R, Edirisinghe M. A novel method for the preparation of starch films and coatings. Carbohydrate Polymers. 2006;**63**:425-431. DOI: 10.1016/j.carbpol.2005.09.018

[59] Espíndola-González A, Martínez-Hernández AL, Fernández-Escobar F, Castaño VM, Brostow W, Datashvili T, et al. Natural-synthetic hybrid polymers developed via electrospinning: The effect of PET in chitosan/starch system. International Journal of Molecular Sciences. 2011;**12**:1908-1920. DOI: 10.3390/ijms12031908

[60] Wang H, Wang W, Jiang S, Jiang S, Zhai L, Qin. Poly (vinyl alcohol)/ oxidized starch fibres via electrospinning technique: Fabrication and characterization. Iranian Polymer Journal. 2011;**20**:551-558

[61] Kong L, Ziegler GR. Quantitative relationship between electrospinning parameters and starch fiber diameter. Carbohydrate Polymers. 2013;**92**: 1416-1422. DOI: 10.1016/j.carbpol.2012.09.026

[62] Kong L, Ziegler GR. Formation of starch-guest inclusion complexes in electrospun starch fibers. Food Hydrocolloids. 2014;**38**:211-219. DOI: 10.1016/j.foodhyd.2013.12.018

[63] Ledezma-Oblea JG, Morales-Sánchez E, Gaytán-Martínez M, Figueroa-Cárdenas JD, Gaona-Sánchez VA. Corn starch nanofilaments obtained by electrospinning. [Nanofilamentos de almidón de maíz obtenidos por electrospinning] Revista Mexicana De Ingeniera Química. 2015;**14**:497-502

[64] Fabra MJ, López-Rubio A, Sentandreu E, Lagaron JM. Development of multilayer corn starch-based food packaging structures containing β-carotene by means of the electro-hydrodynamic processing. Starch/Staerke. 2016;**68**(7-8):603-610. DOI: 10.1002/star.201500154

[65] Hemamalini T, Giri Dev VR. Comprehensive review on electrospinning of starch polymer for biomedical applications. International Journal of Biological Macromolecules. 2018;**106**:712-718. DOI: 10.1016/j.ijbiomac.2017.08.079

[66] Cai J, Zhang D, Zhou R, Zhu R, Fei P, Zhu Z-Z, et al. Hydrophobic interface starch nanofibrous film for food packaging: From bioinspired design to self-cleaning action. Journal of

Agricultural and Food Chemistry. 2021;**69**(17):5067-5075. DOI: 10.1021/acs.jafc.1c00230

[67] Šukyte J, Adomavičiute E, Milašius R. Investigation of the possibility of forming nanofibres with potato starch. Fibres and Textiles in Eastern Europe. 2010;**82**(5):24-27

[68] López-Córdoba A, Estevez-Areco S, Goyanes S. Potato starch-based biocomposites with enhanced thermal, mechanical and barrier properties comprising water-resistant electrospun poly (vinyl alcohol) fibers and yerba mate extract. Carbohydrate Polymers. 2019;**215**:377-387. DOI: 10.1016/j.carbpol.2019.03.105

[69] Mistry P, Chhabra R, Muke S, Sathaye S, Jain R, Dandekar P. Fabrication and characterization of starch-TPU based nanofibers for wound healing applications. Materials Science and Engineering: C. 2020;**119**:111316. DOI: 10.1016/j.msec.2020.111316

[70] Fonseca LM, Radünz M, dos Santos Hackbart HC, da Silva FT, Camargo TM, Bruni GP, et al. Electrospun potato starch nanofibers for thyme essential oil encapsulation: Antioxidant activity and thermal resistance. Journal of the Science of Food and Agriculture. 2020; **100**(11):4263-4271. DOI: 10.1002/jsfa.10468

[71] Rodríguez-Sánchez IJ, Vergara-Villa NF, Clavijo-Grimaldo D, Fuenmayor CA, Zuluaga-Domínguez CM. Ultrathin single and multiple layer electrospun fibrous membranes of polycaprolactone and polysaccharides. Journal of Bioactive and Compatible Polymers. 2020;**35**(4-5): 351-362. DOI: 10.1177/0883911520944422

[72] Alinaqi Z, Khezri A, Rezaeinia H. Sustained release modeling of clove essential oil from the structure of starch-based bio-nanocomposite film reinforced by electrosprayed zein nanoparticles. International Journal of Biological Macromolecules. 2021;**173**:193-202. DOI: 10.1016/j.ijbiomac.2021.01.118

[73] Uygun E, Yildiz E, Sumnu G, Sahin S. Microwave pretreatment for the improvement of physicochemical properties of carob flour and rice starch–based electrospun nanofilms. Food and Bioprocess Technology. 2020;**13**:838-850. DOI: 10.1007/s11947-020-02440-x

[74] Sunthornvarabhas J, Chatakanonda P, Piyachomkwan K, Sriroth K. Electrospun polylactic acid and cassava starch fiber by conjugated solvent technique. Materials Letters. 2011;**65**(6):985-987. DOI: 10.1016/j.matlet.2010.12.038

[75] Sunthornvarabhas J, Chatakanonda P, Piyachomkwan K, Chase GG, Kim H-J, Sriroth K. Physical structure behavior to wettability of electrospun poly (lactic acid)/polysaccharide composite nanofibers. Advanced Composite Materials. 2013;**22**(6):401-409. DOI: 10.1080/09243046.2013.843815

[76] Sutjarittangtham K, Jaiturong P, Intatha U, Pengpat K, Eitssayeam S, Sirithunyalug J. Fabrication of natural tapioca starch fibers by a modified electrospinning technique. Chiang Mai Journal of Science. 2014;**41**(1):213-223

[77] Sutjarittangtham K, Tragoolpua Y, Tunkasiri T, Chantawannakul P, Intatha U, Eitssayeam S. The Preparation of Electrospun Fiber Mats Containing Propolis Extract/CL-CMS for Wound Dressing and Cytotoxicity, Antimicrobial, Anti-Herpes Simplex Virus. Journal of Computational and Theoretical Nanoscience. 2015;**12**(5):804-808. DOI: 10.1166/jctn.2015.3807

[78] Pacheco da CE, Martins FL, Radünz M, Silva da FT, Avila EG, Gandra, da Rosa EZ, Dellinghausen BC. Pinhão coat extract encapsulated in starch ultrafine fibers: Thermal, antioxidant and antimicrobial properties and in vitro biological digestion. Journal of Food Science. 2021;**86**:2886-2897. DOI: 10.1111/1750-3841.15779

[79] Adomavičiute E, Milašius R, Žemaitaitis A, Bendoraitiene J, Leskovšek M, Demšar A. Methods of forming nanofibres from bicomponent PVA/Cationic starch solution. Fibres and Textiles in Eastern Europe. 2009;**74**(3):29-33

[80] Silva I, Gurruchaga M, Goñi I, Fernández-Gutiérrez M, Vázquez B, Román JS. Scaffolds based on hydroxypropyl starch: Processing, morphology, characterization, and biological behavior. Journal of Applied Polymer Science. 2013;**127**(3): 1475-1484. DOI: 10.1002/app.37551

[81] Li S, Kong L, Ziegler GR. Electrospinning of octenylsuccinylated starch-pullulan nanofibers from aqueous dispersions. Carbohydrate Polymers. 2020;**258**:116933. DOI: 10.1016/j.carbpol.2020.116933

[82] Kong L, Ziegler GR. Role of Molecular Entanglements in Starch Fiber Formation by Electrospinning. Biomacromolecules. 2012;**13**(8): 2247-2253. DOI: 10.1021/bm300396j

[83] Martins A, Chung S, Pedro AJ, Sousa RA, Marques AP, Reis RL, et al. Hierarchical starch-based fibrous scaffold for bone tissue engineering applications. Journal of Tissue Engineering and Regenerative Medicine. 2009;**3**(1):37-42. DOI: 10.1111/j.1582-4934.2009.01005.x

[84] Stijnman AC, Bodnar I, Hans TR. Electrospinning of food-grade polysaccharides. Food Hydrocolloids. 2011;**25**(5):1393-1398. DOI: 10.1016/j.foodhyd.2011.01.005

[85] Oktay B, Baştür E, Kayaman-Apohan N, Kahraman MV. Highly porous starch/poly(ethylene-alt-maleic anhydride) composite nanofiber mesh. Polymer Composites. 2013;**34**(8): 1321-1324. DOI: 10.1002/pc.22545

[86] Estevez-Areco S, Lucas G, Roberto C, Silvia G. Active bilayer films based on cassava starch incorporating ZnO nanorods and PVA electrospun mats containing rosemary extract. Food Hydrocolloids. 2020;**108**(106054). DOI: 10.1016/j.foodhyd.2020.106054

[87] Pérez-Masiá R, Lagaron JM, López-Rubio A. Surfactant-aided electrospraying of low molecular weight carbohydrate polymers from aqueous solutions. Carbohydrate Polymers. 2014;**101**:249-255. DOI: 10.1016/j.carbpol.2013.09.032

[88] Valdespino-León M, Calderón-Domínguez G, Salgado-Cruz MP, Rentería-Ortega M, Farrera-Rebollo RR, Morales-Sánchez E, et al. Biodegradable Electrosprayed Pectin Films: An Alternative to Valorize Coffee Mucilage. Waste Biomass Valor. 2021;**12**: 2477-2494. DOI: 10.1007/s12649-020-01194-z

[89] Tuzlakoglu K, Santos MI, Neves N, Reis RL. Design of nano- and microfiber combined scaffolds by electrospinning of collagen onto starch-based fiber meshes: A man-made equivalent of natural extracellular matrix. Tissue Engineering - Part A. 2011;**17**(3-4):463-473. DOI: 10.1089/ten.tea.2010.0178

[90] Rentería-Ortega M, Salgado-Cruz MDLP, Morales-Sánchez E, Alamilla-Beltrán L, Farrera-Rebollo RR, Valdespino LM, et al. Effect of

electrohydrodynamic atomization conditions on morphometric characteristics and mechanical resistance of chia mucilage-alginate particles. CYTA - Journal of Food. 2020;**18**(1):461-471. DOI: 10.1080/19476337.2020.1775706

[91] Gaona-Sánchez VA, Calderón-Domínguez G, Morales-Sánchez E, Moreno-Ruiz LA, Terrés-Rojas E, Salgado-Cruz MDLP, et al. Physicochemical and superficial characterization of a bilayer film of zein and pectin obtained by electrospraying. Journal of Applied Polymer Science. 2021;**138**(12):50045. DOI: 10.1002/app.50045

[92] Rentería-Ortega M, Salgado-Cruz MDLP, Morales-Sánchez E, Alamilla-Beltrán L, Valdespino-León M, Calderón-Domínguez G. Glucose oxidase release of stressed chia mucilage-sodium alginate capsules prepared by electrospraying. Journal of Food Processing and Preservation. 2021;**45**(5):e15484. DOI: 10.1111/jfpp.15484

Chapter 10

Micro and Nano Structuring as Method to Enhance the Functional Properties of Starch-Based Polymeric Materials

Doina Dimonie, Mircea Filipescu, Mihai Dragne, Alina Mustatea and Nicoleta Dragomir

Abstract

The use of starch, the second most abundant natural resource in the word, as polymer is unprofitable and limited by certain functional properties. The structuring of multiphase polymeric materials represents the process of diminishing the dispersed phases till micro-and/or nano-dimensions and the positioning of the resulted fields in an order through which the properties of interest are achieved as far as possible. The structuring is reached by controlling the interface properties for achieving physical, chemical, biological or rheological compatibilization, mainly by melt compounding procedure. The chapter proves that the structuring of starch-based multiphases polymeric systems by reactive compatibilization is a good possibility to guarantee the functional properties, required by sustainable applications, of interest even in 2050 perspective. The chapter underlines also that the structuring by reactive compatibilization is connected with the formulation designing and choosing of the melt-compounding conditions in such a manner for the chemical bonding of the minority phases with the main polymeric matrix and therefore increasing the component miscibility and the functional properties of the resulted materials till the requirements of the sustainable applications.

Keywords: starch, multiphase polymeric systems, interface, reactive compatibilization, reactive extrusion

1. Introduction

After cellulose, starch is the second most abundantly available natural polymer in the world. Because of its biodegradability and non-toxicity to the natural environment, it becomes a raw material very attractive for the food and non-food application. Starch is a homo-polysaccharide made up of glucose units, linked together via glycosidic linkages, with a renewable-botanic origin (seeds and plant tubers), lowest price and highest industrialization potential for the next decade [1–13]. Starch has a biphasic composition because consists in two polysaccharides, amylose with linear macromolecules and low molecular weight and amylopectin with branched, clustered chains which belong simultaneously to several clusteres and high molecular weight. Depending on its origin, the diameter of the starch granules varies between 1 and 2 μm to 100 μm. The starch granules are different shaped: round (corn), oval (potato), rounded and truncated at one end (tapioca), flat, elliptical (wheat) [4–9]. Because each starch is different in composition the obtained films could exhibit different properties [10].

The ratio between the two constituent polymers of starch conditions the subsequent use of starch as thermoplastic polymer, as during the melted state flowing, the amylose macromolecules align in the flow direction and because its chemical strongly branched macromolecules, those of amylopectin cannot be aligned. The ratio between amylose and amylopectin depends on the starch origin and the growing climatic conditions. Starch is a semi-crystalline polymer that does not melt in the traditional sense to form a liquid. Starch melting occurs in the presence of a moderate (10–30%·w/w) water content. Starch crystals contain about 9–10%·w/w of bound water, which does not freeze at cooling below 0°C. Additional water or plasticizers is required for melting of starch at convenient temperatures below the water boiling temperature and the starch degradation temperature [11]. Starch has multiple glass transitions under which physical aging occurs. Due to the high hydroxyl content, the starch properties are strongly dependent on the moisture.

Currently, on the market, there are varieties of commercially starch available (potato, corn, wheat, topioca, etc)and numbers of un-explored and under-utilization sources of starch (fruits processing waste, different agro-industrial residues, etc.) [12, 13].

Starch is an important source of raw materials both as organic compounds and polymer used for plastic industry. It can be converted readily into a variety of useful monomeric and polymeric products by chemical and biochemical routs. Fermentation of starch to industrial-grade ethyl alcohols becoming more economically competitive with the synthetic methods due, largely, to the increasing cost of ethylene. Other biological conversions yield a variety of alcohols, ketones, and organic acids, and these too will become increasingly important as the petroleum situation worsens. Enzymatic conversion of starch to glucose, a useful starting compound for a variety of alcohols, acids, and polyols, proceeds readily and in near quantitative yield. Glycol glucosides, cyclic polyols derived by glycolysis of starch, are excellent replacements for petroleum-derived polyols in the production of rigid urethane foams, and can replace up to 85% of petroleum-based polyols in alkyd resins with no loss in quality of the resin [14].

However, its use as a polymer is unprofitable and limited by certain functional properties. Its hydrophilicity, thermal, and mechanical properties limitations, low physical properties (brittle, poor moisture resistance, low permeability to lower gases, high density) rapid degradability, and strong intra and intermolecular hydrogen bonding of the polymer chains hinder its melt processability and limit its widespread commercial application as a renewable biopolymer [3, 15]. Starch also is easy degradable under flowing in the melted state. Moreover there are many difficulties in controlling the functional properties during its service life because of frequently occurred phenomena as exudation of plasticizers (anti-plasticization), re-crystallization (retrogradation) [7].

The perspective for polymeric materials based on renewable resources will be constantly increasing as far as 2050. Due to the failure in solving the issue of environment infestation with secondary polymeric materials, the development policies in designing and achieving polymers and materials based on them are mainly oriented towards sustainable applications.

In order to reach functional properties of practical interest including the sustainable applications, the multiphase starch-based systems are structured mainly by reactive compatibilization at melt processing (reactive extrusion) which has proven to be the most effective method. That is why the chapter presents the possibilities to get new starch-base, multiphase materials proving that structuring based on reactive compatibilization are adequate and practical solutions to ensure functional properties for durable applications. The chapter presents the main possibilities of structuring multiphase systems based on starch and details certain aspects related to the compounds of starch with polyvinyl alcohol.

2. Structuring the multiphase polymeric materials

Multiphase polymeric materials are homogeneous or non-homogeneous systems which contain solid, liquid or gaseous phases and include categories as: composites, blends, gels, interpenetrated polymer networks, mono or layered structure, cellular solids (foams), including biological type (hydrogels) etc. [16, 17]. The dispersed phases into the polymeric matrix can be functionalized, microencapsulated, oriented, continuous or discontinuous, etc. [16–19].

The structuring of multiphase materials represents the process of diminishing the dispersed phases till micro-and/or nano-metric scale and the positioning of the resulted mico and/or nano fields in an order in which, the properties of interest are achieved as far as possible. The structuring consists in enhancing by controlling the interface properties the physical-chemical-biological-rheological compatibility of the blend components considering methods connected with mobility of the segments/ macromolecules/morphological elements, or with reactivity of the components or/ and with ordering of the flowing elements at melt compounding [16–18].

The structure-properties relationship has a crucial significance in the structuring of the multiphase polymeric materials for developing grades satisfying varied engineering requirements [18, 19]. With the help of modern technologies, the circumstances in which the phases of the heterogeneous polymeric materials can be so arranged to give the desired properties in correlation with the intended applications, is possible to be identify.

The interfaces adhesion between the dispersed phases and polymeric matrix are maintained by intermolecular physical or chemical forces and/or by chain entanglements [20]. In the absence of these interactions, the interface becomes the place where the brittle fracture takes place. The interface can be seen as a resistance between materials in contact and is characterized by interfacial tension (controlable by changing of the Gibbs potential per unit area). The interfacial tension depends on factors such as: phases geometry (shape, average size, particle size distribution, porosity, etc.), the characteristics of macromolecular chains (molecular structure, molecular architecture, tacticity, crystallinity, ramifications and/or, defects, impurities, ash), the size of thermo-mechanical stress, the melt flow features (the size of mechanical stress, temperature) [16, 21–25]. Most thermodynamic studies of multiphase polymer systems are based on Flory-Huggins' theory [26].

Changing the interface properties have the effect the modification of the material properties. The least common possibilities encountered in the practice of polymer blends achieving are those in which the compound properties are synergistic, i.e. the blend properties are greater than the additive properties of the individual polymers. The properties are synergistic for polymers showing high interface adhesion denoting that they are thermodynamically compatible [17, 27, 28]. The most common dependence blends properties - composition is the additive those, when the blend properties represent the average of the component properties in their pure state, without minimum or maximum. If the interfacial adhesion is very weak and the polymers are thermodynamically incompatible, this dependence does not follow the law of additivity, but shows a variation with a minimum, which means that the blend properties are lower than those of each individual components [17, 28].

Compatibilization is the process of modifying the properties of the interfaces in immiscible polymeric systems which results in the creation of the interphase region in which a gradient of material composition is created [20]. By compatibilization the morphology of the new material is stabilized at macroscopic level and therefore, the functional properties are constant over lifetime. The degree of compatibility is estimated by the thickness of the interface, the size of the dispersed phases, the mechanical performances generally by all functional properties. Unlike the compatibility, the miscibility is related to the blending of components at the molecular level; in the whole mass, which means that in the case of miscible blends it is no longer about dispersed phases [20, 29].

In order to improve the compatibility, the problem of minority components dispersing into the majority polymeric matrix is solved by reducing the interfacial tensions using one of the following techniques: entanglement of the macromolecules, interpenetration of interfaces, development of chemical (covalent linkages) or physical (e.g.van der Waaals, hydrogen bonds, ionic interactions, etc.) bonds [22]. The entangled interfaces are found in miscible blends or composites with good interfacial adhesion. The interfacial tension can be controlled by using polymeric emulsifiers (grafted or block copolymers, others) which create secondary bonds or by using compatibilizers, which generate chemical bonds between phases, after reactions possible between several functional groups (anhydride-amine, epoxy-anhydride, oxazoline-carboxylic acid, isocyanate-carboxylic acid, lactic acid-amine, carbodiimides-carboxylic acid, ion exchange reactions, free radical grafting reactions at the phase interface [17]. By using a controlled method to improve the compatibility even miscibility between the components of a new multiphase polymeric materials is possible to achieve materials, with own functional properties correlated with requirements of various applications.

In case of multiphase polymeric materials with target fillers, the improving of compatibility is based also on the creating of physical and/or chemical bonds. If the cohesive energy thus developed is below a critical value, then morphological defects are created, between the filler and matrix, representing space filled with air, usually found as micro-voids, voids, macro-voids, cracks, fractures. The characteristics of the empty space depend on the shape of the filler. The cohesive energy may be below a critical value, situation in which the miscibility between the components is weak and/ or when the distribution of the filler into the matrix is inadequate. In the engineering practice, there are no free-defects multiphase materials such as the mentioned those it was found that if the content of these defects is higher than 5%, then the material functional properties decrease with approx. 30%. The structural defects presence is dangerous because, under stress, it accelerates the mechanical destruction by cracking and breaking [30, 31]. A filler can act also as compatibilizer when the radius of the particle has the same order of size as the radius of gyration [32].

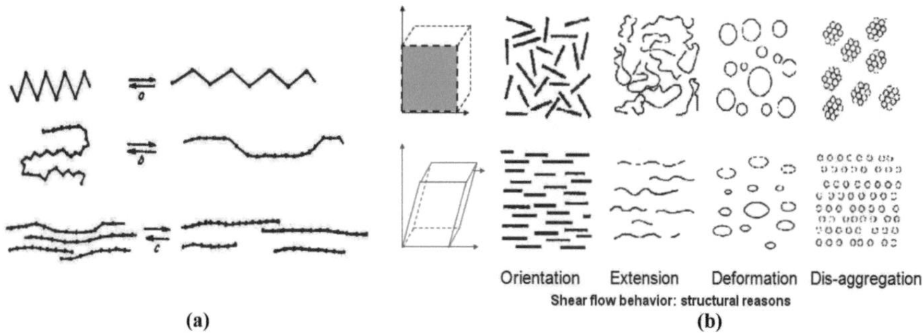

Figure 1.
Structuring melts by flowing in elongational (a) or shear stress field (b) [17, 33].

Changing the interface properties has as effect the adjusting of the polymeric system properties. The least common possibilities encountered in the preparation of polymer blends are those in which the dependence of compound property-composition is "synergistic", i.e. the blend properties are greater than the additive values of the properties of individual polymers. The properties are synergistic only in case of the polymers showing high interface adhesion which means they are thermodynamically compatible [17, 27, 28]. The most common dependence blends properties—composition is the "additive" those, when the properties of the blend represent the average of the properties of its components in their pure state, without minimum or maximum. If the interfacial adhesion is very weak and the polymers are thermodynamically "incompatible", this dependence does not follow the law of additivity, but shows a variation with a minimum, denoting that the blend properties are lower than those of each individual components [17, 28].

The quality of items obtained by melt processing techniques, expressed as morphological uniformity, anisotropy of properties, appearance of surfaces, etc. depends on the structuring under stress of the flow entities in the field of mechanical and thermal stresses. Structuring under stress is the result of the molecular disaggregation, deformation, orientation, extension of macromolecular chains developed when the melts in passing through the nozzle of the nozzle of the molder device (**Figure 1**) [3, 33].

The flow resistance of the polymeric melts, a parameter which control the structuring during the melt compounding flowing, depends both on the shear rates and the molecular parameters describing the chemical structure of the polymeric blends components [34]. As an effect of the order degree gained during the melts flow through the nozzles is the crystallization under stress [33].

In order to deepen the structuring phenomena, the elucidation of the correlation between the molecular parameters (degree of polymerization, Flory-Huggins interaction parameters, etc.) and macroscopic phase separation phenomena is necessary to be done by investigating the heterogeneous structure from submicron scale to nanometer those [35].

3. Structuring of starch based materials

To get functional properties of wide applicability, starch can be converted into multiphase polymeric compounds, by physical and / or physical-chemical and/or chemical modification with other polymers and/or non-polymeric materials [36] considering un-reactive and/or reactive melt compounding [25, 37–40]. Generally speaking, starch has low compatibility with other polymers or biopolymers and therefore the degree of compatibility varies depending on the specific used grades. The success of the starch converting into such multiphase polymeric systems is conditioned by the achievement of the molecular miscibility or, at least, by the creation of an advanced dispersion of minority components into the main polymeric matrix, via lowering the interfacial tensions considering the described compatibilization techniques [24, 41–43].

The convertion of starch into multiphases polymeric compounds of practical interest must consider both the starch origin, the ratio between the amylose and amylopectin, the constancy of this ration (especially for scaled up procedures), the correlation of the modification method with all the starch structural characteristics and the requirements of the considered application. The structuring possibilities are related equally to the way of designing the formulation and choosing of the melt compounding conditions the chapter presenting examples regarding the starch-polyvinyl alcohol compounds. An overview of the reactive compatibilization as method to get structured multiphase materials based on starch with other polymers is also given. A special attention is paid to the possibilities of avoiding the melt degradability of the multiphase systems based on starch. The chapter details also the structuring methods to achieve starch - PVOH nanocomposites with layer silicate content and presents a silicate treatment method which leads to the formation of exfoliated nanocomposites. The structuring through incorporation of a gaseous phase into the starch-starch matrix is also shortly described.

4. Structuring through melt compounding

Starch can be compounded with renewable polymers from *algal origin* (alginates); *microbial origin* (polysaccharides: chitosan, curdlan, gellan, pullulan, xanthan, bacterial cellulose; polyesters and copolyesters: polyhydroxyalcanoates (PHA), poly (butylene adipate-Co-terephthalate) (PBAT)); *plant origin* (polysaccharides: cellulose, pectin; proteins: soy proteins, zein, gluten: polyesters (polylactic acid PLA)); *animal origin* (polysaccharides: chitin, chitosan) [44, 45]. On the market there are also of interest starch compounds biodegradable or water- soluble oil-based polymers as: polycaprolactones (PCL), polybutylene succinate (PBS), polybutylene succinate adipate (PBSA) or polyvinyl alcohol (PVOH) [46].

Starch compounds with PVOH are designed for replacement of LDPE films in application where barrier properties are not critical, water-soluble laundry bags, biomedical and clinical fields, replacement of polystyrene foams as loose fill packaging materials, packaging applications. Blending of starch with other biodegradable polymers such as biodegradable polyesters; polylactic acid (PLA) and polycaprolactone (PCL), was recognized as a successful strategy to provide a renewable, fully biodegradable and cost-effective materials [47]. The compounds of starch with PLA are conceived for applications as food packaging, electronic devices, membrane materials (chemical and automotive industries), medical applications, packaging materials. The compounds of starch with PBSA are useful as antimicrobial packaging materials and those with PHB as biomaterial in medical applications. PBAT starch compounds have controllable mechanical properties and are therefore are designed for multiple types of packaging [48]. The use of starch in achieving of polyurethane (PU) multiphase materials not only yields PUs with outstanding mechanical properties but also makes the final PU products biodegradable. The hydrophilic nature of starch limits its dispersion in hydrophobic PU polymers [49] and therefore these compounds must be reactively compatibilized.

Renewable fillers play a valuable role in the development, of new multiphase polymer materials based on starch considering melt compounding techniques. It should be noted that the degree of compatibility between starch and other biopolymers extensively varies depending on the specific biopolymer, generally starch having as has been said low compatibility with other polymers [43, 50].

4.1 Melt compounding conditions

Starch can be melt compounded considering various conventional processing techniques, similar to those widely used for typical synthetic thermoplastics, such as extrusion, injection, compression molding, casting and foaming, as well as some new techniques as melt reactive melt compounding. Various starch-based products which have been developed and commercialized as extruded films/sheets, foams, shaped articles, etc. [51] using internal mixer, mono (**Figure 2**) or twin-screw extruders, injection devices, others can be obtain [52]. Formulations often require inlets for plasticizers, filler or other additives. Escape of volatiles such as steam will be required, without loss of other materials. High shear is required to disrupt the native starch structure and produce a uniform composition with other components. The extrudate must be a uniform continuous stream with rheology suitable for shaping [11].

A twin-screw extruder has a large operational flexibility (individual barrel zone temperature control, multiple feeding/injection, and screw configuration for different degree of mixing/kneading) and is useful for intensive mixing and compounding of components into starch plastics. In a twin-screw extruder the custom combinations of rheological elements can be assembled along the screw. Zone of high shear will assist with disruption of granules while uncoiling of molecules can take place in less shear intensive zones. Another advantage of the twin-screw extruder is to allow the decoupling of die flow and mechanical treatment. During extrusion of starch-based multiphase materials, residence times and specific mechanical energy inputs must be controlled, and high efficiency production can be achieved [52]. The temperature conditions for preparation and processing of these materials must be chosen to minimize possible degradation of the organic modifier and the matrix [53–55]. The shear experienced during compounding may have caused fragmentation of starch and destroyed its crystalline structure [56]. Shearing of the molten granules destroys their organized structure, and crystalline, granular starch is converted to a dispersed, essentially amorphous material [57]. It was found that if the extrusion is performed at

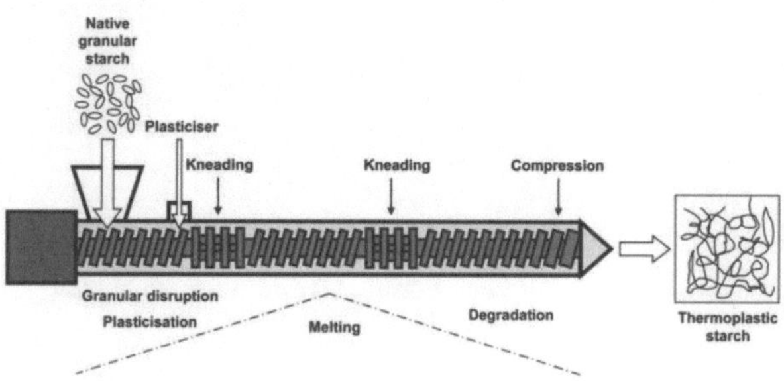

Figure 2.
Schematic representation of starch processing by extrusion [52].

high rotational speed and high stationary times the single helix crystallites are formed [58, 59]. Also during extrusion the shear stress may enhance starch separation, whereby amylose may be partly leached out of the amylopectin [60, 61].

Because the melt compounding conditions control the melt rheological properties they have a great impact on the properties of the resulted compounds [42, 52]. In [42] it has been shown that for the polymeric compounds sensitive to the melt processing parameters, such as those based on starch, the boundary between miscible and less miscible is very small and therefore each compounding ratio requires its own formulation and its own melt flow conditions. It was shown that the miscibility of starch-based compounds increases if, for each formulation is found the melt processing conditions which ensure the better improving of miscibility because the melt resistance to flow allows a suitable homogenizing without degradation of the macromolecules (**Figure 3**).

The starch-PVOH compounds designed for short life application which had enhanced miscibility achieved by controlling the melt resistance to flow via melt compounding conditions have a better surface appearance (**Figure 4**) and no longer presents the anti-plasticization and retrogradation phenomena [42].

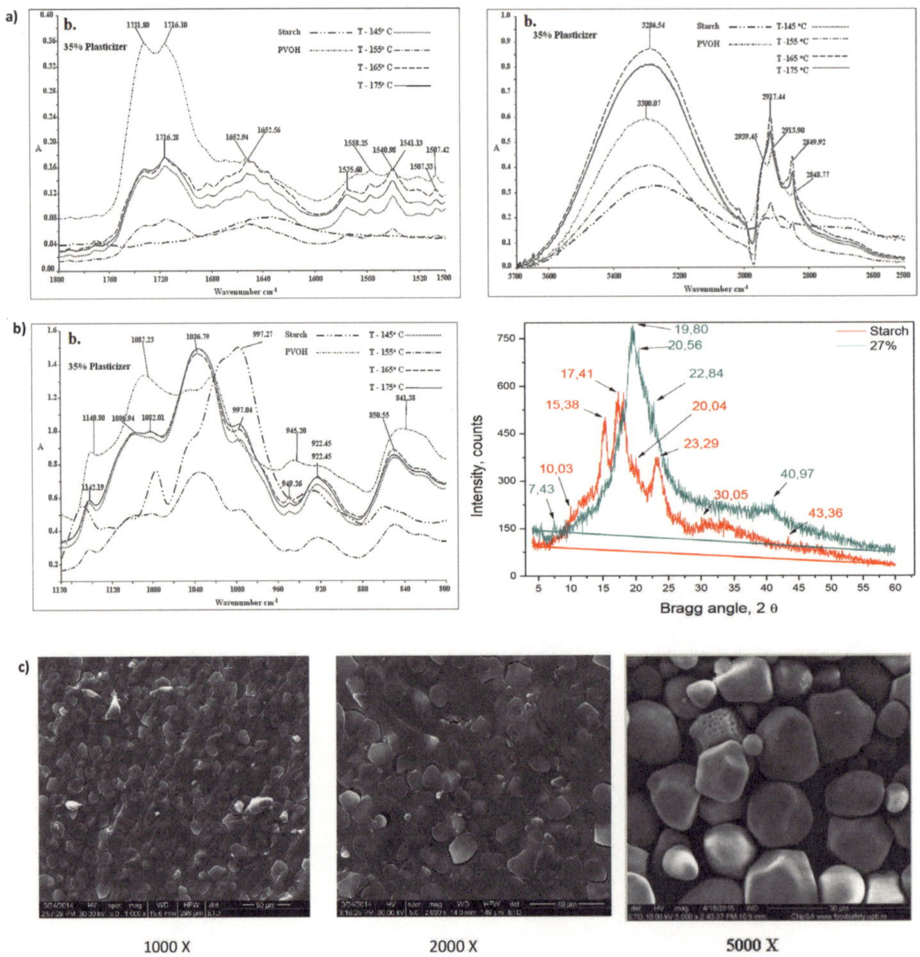

Figure 3.
Dependence of the shape of the FTIR spectra (a) and of the XRD diffractograms (b) and the SEM micrographs aspect (fracture) of some starch-PVOH compounds (c) on the melt compounding conditions [42].

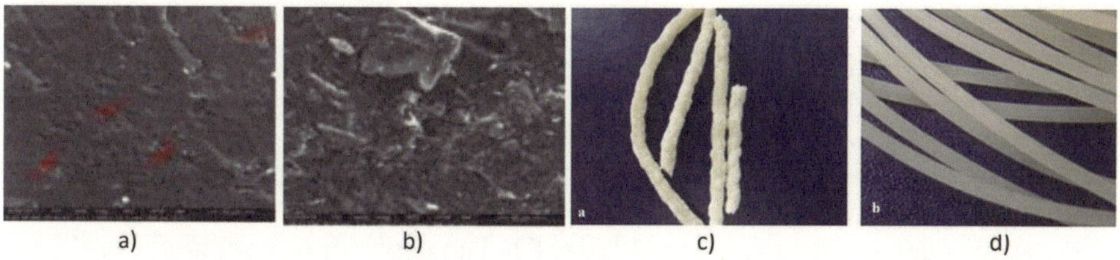

Figure 4.
Morphology (fracture, 1000×) (a, b) and surface appearance (c, d) before (c) and after (d) improving the miscibility via controlling the melt resistance to flow of some starch-PVOH compounds [42].

Figure 5.
Granules (20–70% starch content) and items achieved from PVOH-starch compounds after elimination of defects via structuring by controlling the melt resistance to flow [17, 63].

It was observed that the starch polymer melts exhibit strong elastic properties and therefore phenomena as nozzle swelling, shark skin and wall slip [62] can be controlled. The experimental results have shown that these phenomena can be avoided by controlling the elastic component of the melt vasco-elastic properties considering the formulation and the selection of optimal compounding conditions. In this way the PVOH-starch compounds can be melt processed into items without defects and degradation, with natural color, smooth surface (**Figure 5**) and functional properties of interest for short-life applications [17, 63].

5. Solid state properties conditioning the structuring efficiency

5.1 The polymers particles shape and size influence

In [64] was found, that the turning of corn starch via melt compounding with PVOH can be achieved by providing similar melting behavior of the starch and of PVOH powder and by controlling the process sensitivity of the new compounds. The corn starch and PVOH particles had their own size distribution and shape. If the starch particles were rectangular, those of PVOH were spherical [64]. The PVOH particles shape depends on the manufacturing technology and can be spherical only if the polymer was obtained by reverse suspension hydrolysis of poly (vinyl acetate) (PVA) [65]. It has been shown that the particles of the two polymers cannot be melt compounded than after the selection of fractions with particle size variable the same range. If the particle size distribution of the two polymers is wide, then the obtained compounds include defects, either rough surfaces or un-melted inclusions. Variable sized particles have distinct melting times, longer for those with larger diameter and smaller for those with smaller diameter. The rough surface results due to the degradation of small particles and solid inclusions represent un-melted particles because they had large diameter and have no time to melt. Due to the variation of the particle sizes of the two polymers even the morphologies of the achieved compounds and their dynamo-mechanical properties are affected (**Figures 6** and **7**). If the particles of the two polymers have approximately the same size then they melt in approximate identical time and the two type of defects are avoided.

In addition, if the blends obtained through melt compounding of starch with PVOH contains un-melted particles, then morphological defects such as voids, cracks fractures, may appear around them (**Figure 6**), The new compounds always will contain, near un-melted particles because of their size, parts from the starch grains, representing the branched macromolecules of amylopectin, very well visible on the SEM micrographs.

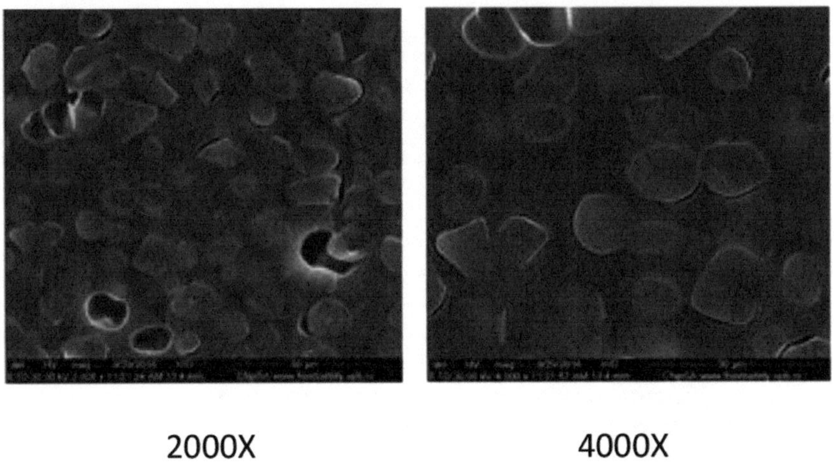

Figure 6.
Morphological defects of the compound achieved from starch with 40–63 μm sized particles and PVOH with particles of 80–100 μm (1/1 blending ratio) [64].

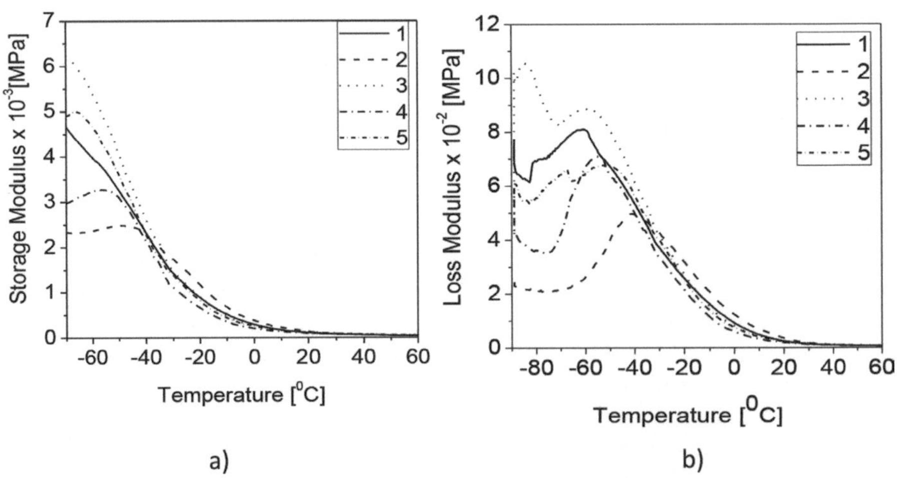

Figure 7.
The dependence of the dynamo-mechanical properties of the starch-PVOH compounds (storage modulus (a), loss modulus (b)) on the temperature and the particle size (1—P1 (80–100 μm)/P2 (80–100 μm); 2—P1 (80–100 μm)/P2 (100–200 μm); 3—P1 (80–100 μm)/P2 (200–300 μm); 4—P1 (63–80 μm)/P2 (80–100 μm); 5—P1 (63–80 μm)/P2 (80–100 μm)) [64].

Considering the word "compatibility" describing the partial miscibility of the compounds components which result in macro-metric scale time stability of the obtained compounds, appropriate for the life-time, the term "miscibility" reflects the interpenetration of the components at the molecular level and so stability for an unlimited period for the new materials [64]. It can be appreciated that the polymeric particles, through their size, make the differences even between compatible and miscible compounds. In analyzing the compatibility/miscibility between starch and PVOH compounds (claimed compatible by some authors and opposite by others) the influence of the particles size and of the amylopectin content should be equally considered. In accordance with [64] the starch-PVOH compounds obtained from small sized particles are close to be totally miscible, presenting a single glass transition without shoulders, totally different as those of starch and PVOH beeing almost a continuous phase, only with few defects as voids etc. In the last period, on the commercial market, can be found grades of thermoplastic starches which represent plasticized starch powder melt processed into granules.

6. Structuration through reactive compounding (*in situ* compatibilization)

The experimental practice has shown that only physical compatibilization which establishes only secondary bonds between the dispersed phases and the matrix does not generate functional properties of practical interest. As it has been pointed out, because of poor interactions, starch is generally thermodynamically immiscible with all renewable polymers showing a dispersed structure (ex. ternary blends based on PCL/starch/PLA, starch-PBAT [66]. Structuration through reactive compounding (*in situ* compatibilization or chemical compatibilization) involves the development, in the melted state, during compounding, of chemical reaction between the functional groups of the compatibilizer and those of the starch and the second polymer from the compounds. The process involves the *in-situ* formation of amphiphilic structures, and thus covalently bonding the phases in the melt state. Reactive melt compounding is a convenient and cost-effective technique for the esterification of polysaccharides using anhydride reagents In this way the interface tension decreases, the size of the dispersed phases decreases and the functional properties can be controlled [67, 68]. The chemical compatibilization is relevant when morphology present regions with fully continuity or presents a co-continuous aspect.

To understand the lack of affinity between the different phases, interface adhesion can be investigated by contact angle measurements. The compound morphology can be better observed using microscopy techniques (e.g. SEM, etc.) The rheological investigations in solid (DMA) and melted state helps to understand better the blend structure formation during the process [69]. Currently, nanofillers are used to improve both mechanical properties and the phase morphologies of immiscible blends. The main advantage of using nanofillers, when compared to copolymers, is that the former can simultaneously act as nano-reinforcements and compatibilizers [67]. There are situations when the reactive compatibilizers can fulfill other functions such as antibacterial agent [70].

Twin-screw extruders are typically used for reactive compounding due to their excellent control of mixing, temperature, and residence time distribution, with highly accurate feeders, systems for removal of reaction heat, etc. [71–75].

Because of interest for industrial quantities of compatibilized multiphase renewable polymeric materials and the possibility to ensure functional properties of practical interest for desired applications, the chemical compatibilization strategy is usually implemented in reactive melt processing procedure (reactive extrusion) [71]. The concrete reactive compatibilization solution depends on the nature of the renewable multiphase polymeric system which must be structured [76–78].

7. Structuring by incorporating layered silicates

Polymeric nanocomposites have achieved much more attention due to their enhanced physic-chemical and mechanical properties, improved moisture sensitivity in comparison to the pure polymers. Due to its chemical and physical properties, starch is highly valuable to be structured through converting into nano-composites considering melt compounding procedures (twin screw extruder) including the reactive melt processing [79, 80].

Various types of nano-fillers that have been used with plasticized starch can be used such as montmorillonite, cellulose nano whiskers, cellulose nano-fiber, and starch nano-particles (obtained by acidic hydrolysis of waxy maize starch granules as reinforcement [81]). The starch-based nano-composites with different nanofillers for reaching a optimal dispersion and properties need particular preparation strategy [82–85]. The incorporation of various fillers in a starch-based polymer matrix generates specific structuring process because in this situation they control the specific orientation and/or crystallization processes.

For getting starch-based nano-composites, blends of starch with bio-polyesters or other renewable- or synthetic-based polymers can be used, e.g. which can be considering various obtaining procedures [86].

Regarding the achieving of the starch-PVOH nano-composites structured with layered silicates in a melt compounding procedure were studied aspects as: the dependence of miscibility, morphology and other functional properties on the layered silicate nature, the functionalization strategy of the galleries [87, 88] the target filler content [89], the incorporating method of the layered silicate into the starch matrix [90–92], the correlation between miscibility and the surface defects [31, 93]. At first sight, the dispersion of the layered silicate with the help of shear stress at melt compounding should favor the exfoliation process. In reality, this method is limited both by the low thermo-stability of the polymers and those of the used modifiers but also due to the loss of hydration water of the layered silicate consequence of long maintenance at high temperature in the compounding device. It was also found that, at the same target filler content, the size of the dispersed phase and the number of dispersed phases per unit area (drops) depend on the type of filler, nature of the surface treatment and particle size. It was concluded that the degree of intercalation of the layered silicate with the PVOH-starch matrix increases if purified layered silicate (NaMMT) or ammonium ion functionalized layered silicate (Nanocor I 28) are used to strengthen the STARCH-PVOH matrix. If, on a macroscopic scale (100 μ) the starch-PVOH-layered silicate compounds seem to be compact materials, at microscopic level (10 μ), defects as gaps, cracks, fracture etc. appear. These morphological defects seem to illustrate a disordered microstructure, due to the poor distribution of swollen silicate tectoids into the continuous polymeric phase, possibly due to inadequate compounding conditions. It has been shown that the PVOH-starch-treated layered

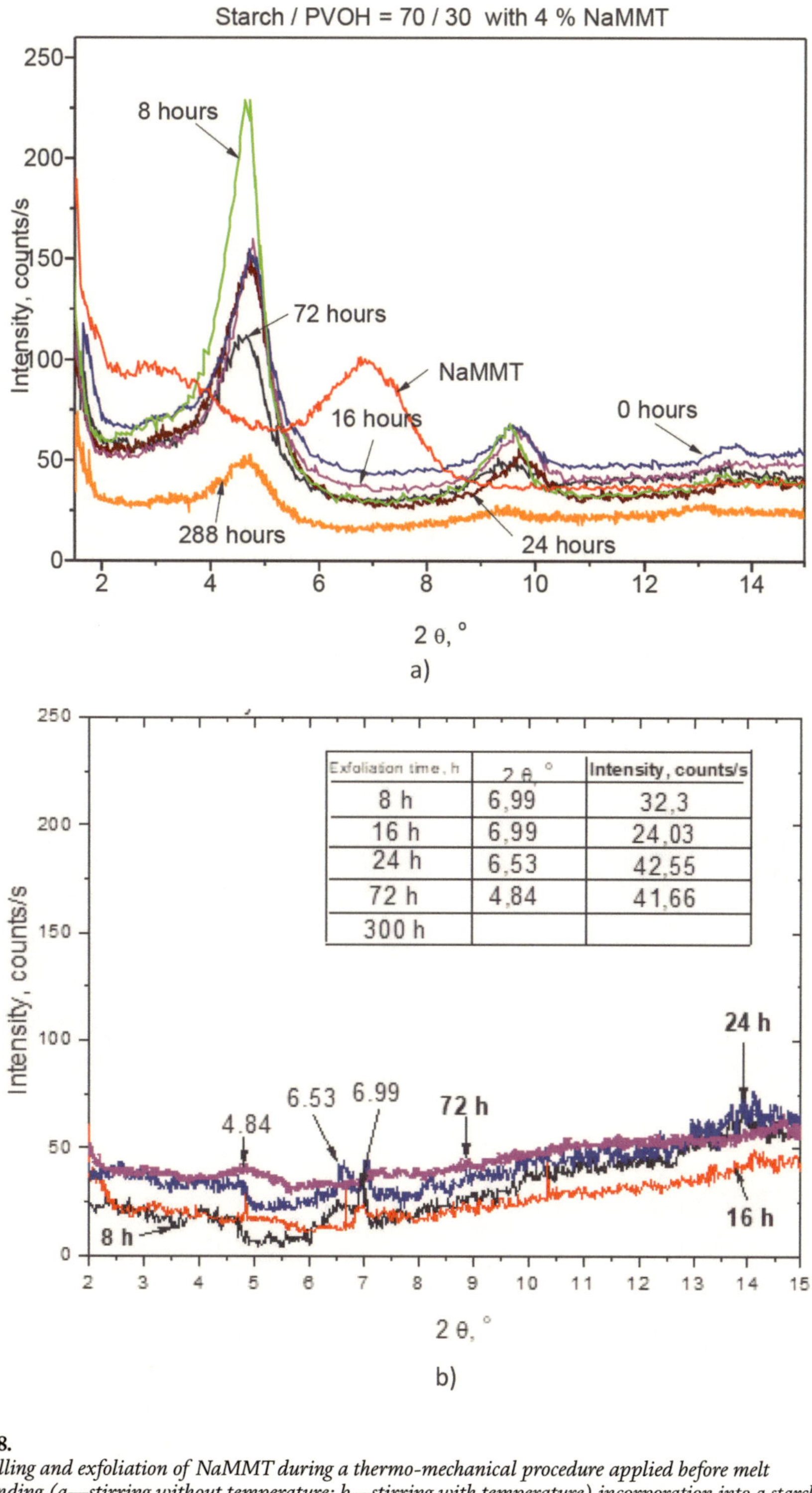

Figure 8.
The swelling and exfoliation of NaMMT during a thermo-mechanical procedure applied before melt compounding (a—stirring without temperature; b—stirring with temperature) incorporation into a starch-PVOH matrix [91].

silicate blends which, according to X-ray diffractograms, appear to be of exfoliated type and which have an improved thermal behavior, including in terms of uniformity of the melting process, have homogeneous surfaces and low material imperfections as gaps, cracks. SEM micrographs show that the surface of the compositions which, according to X-ray diffraction, are exfoliated intercalated nano-composites. Are characterized by small contact angle, are nano-structured, show advanced miscibility (demonstrated by FTIR analysis), have smaller number of defects as gaps, cracks, fractures [31, 87–93]. Compounds that, according to XRD diffractograms, are micro-structured, have frequent surface defects and irregular fracture areas, which are the starting points for cracking, crack propagation and stress rupture [31, 87–93].

In order to achieve a good swelling of the layered silicate and a homogeneous dispersion of the obtained lamellae into a starch-PVOH compound, it has been shown that, the layered silicate needs to be swelled, better in a pre-compounding procedure [91]. The layered silicate (NaMMT) was treated, before compounding in water, an effective starch and PVOH plasticizer. It was shown that the degree of exfoliation of the multi-layer silicate and the properties of the achieved micro- and nano-composites depend on the way the silicate was treated (stirring of layered silicate at 1500 rpm, in water, at room temperature or at 50°C, for (8, 16, 24, 76, 288 h) (**Figure 8**). A small degree of silicate exfoliation reveals that the layered silicate was not sufficiently hydrated and therefore did not disperse well in water. If the silicate was treated an optimal time at room temperature or better at 50°C, a much shorter time, then the XRD diffractograms show the swelling of the NaMMT lamellae and even their exfoliation [91]. The obtained

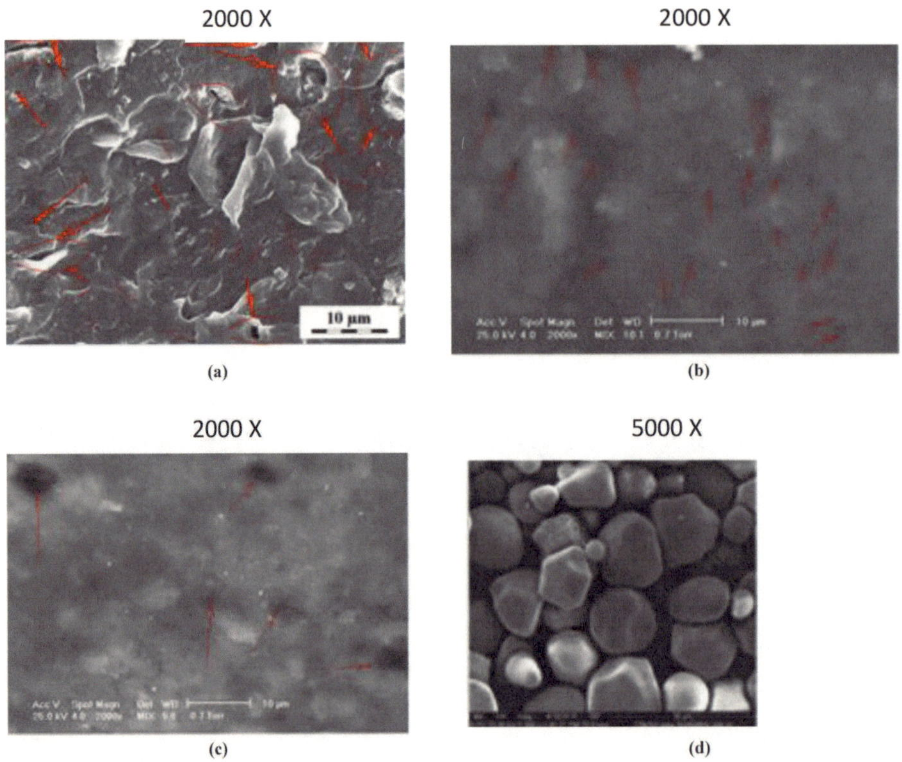

Figure 9.
Influence of silicate treatment on the surface appearance of a multiphase compound with 70% starch and 4% NaMMT (untreated (a), treated at 50°C, for 8 h (b) and treated for 24 h (c); starch particles (d) (SEM micrographs).

results show that even for starch-PVOH compounds the target fillers, well selected and properly embedded, function as efficient interface agents.

If the surface appearance of the PVOH-starch composites containing untreated NaMMT is compared with that which enclosed pretreated silicate at 50°C for 72 h, reveals that for the latter, the surface defects and the included spherical shapes with well-defined interfaces are almost disappeared (see SEM micrographs from (**Figure 9**).

As consequence of the layered silicate de-lamination the storage modulus increases from 2 to 6 MPa, for the starch-PVOH compound without silicate, to 35–40 MPa, if the blend contains NaMMT stirred with water at 50°C, for 72 h. Depending on the characteristics of the applied treatment, starch-PVOH based micro- and/or nano-structured, intercalated and/or exfoliated nano-composites were obtained [91].

8. Structuring by incorporating gaseous phase

Depending on the dynamo-mechanical modules and the glass transitions, starch-PVOH formulations were selected for obtaining biodegradable structured foams of practical interest, with a compressive strength of 5–25 MPa and which supports compressions of 80–90%.

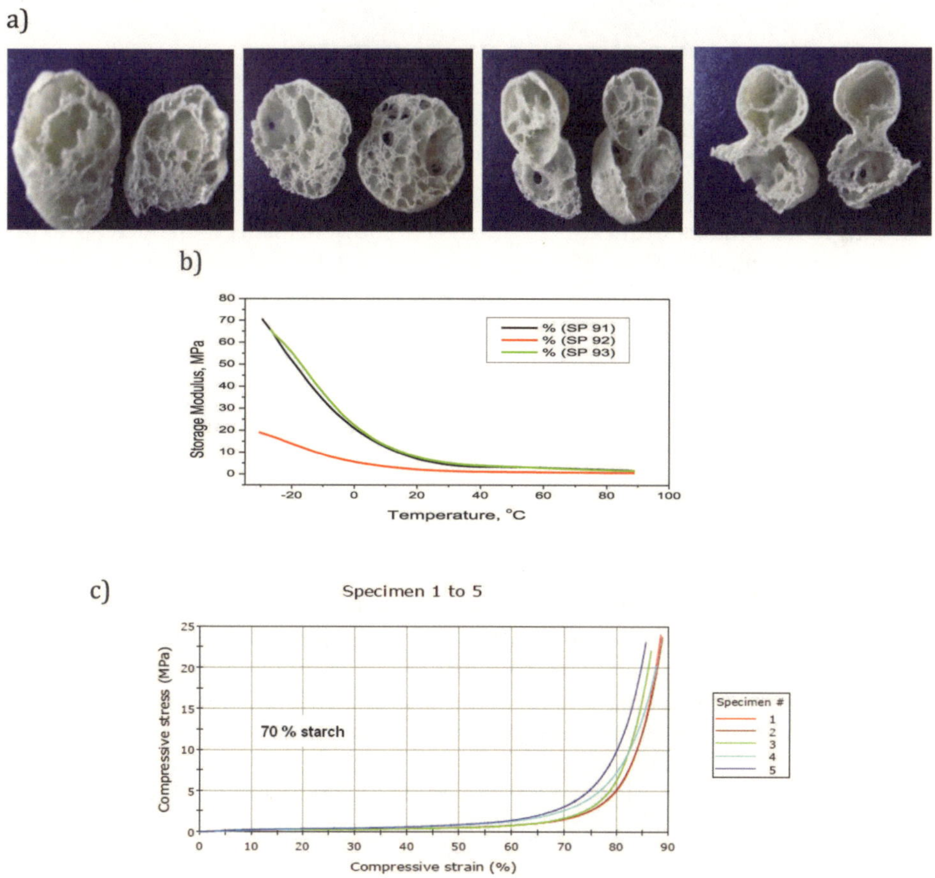

Figure 10.
Starch-PVOH-based foams (a) and their properties; (b) storage modulus; (c) compressive stress [17].

Property	Starch content, %			
	10	20	30	40
Storage modus, T = −30°C, MPa	4.5	3.5	2.5	2
Loss modulus, T = −30°C, MPa	6	7	5	4.4

Table 1.
Dependence of the storage and loss modulus on the starch content of the basic blend [17].

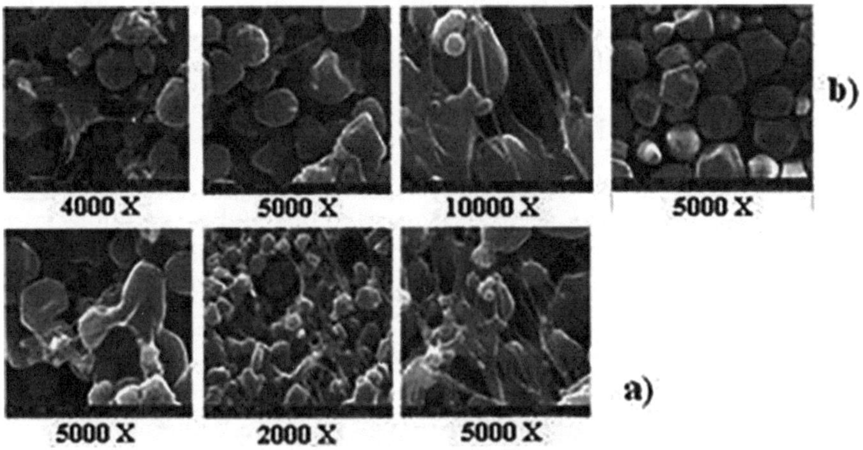

Figure 11.
Morphology of starch-PVOH based foams (a); shape and size of the starch particles (SEM micrographs) (b) [17].

Biodegradable foams with variable starch content were made, with open pores of 2–3 mm, storage modulus 2–40 MPa, loss modulus of 2–8 MPa and which supports a compression deformation of 60–85% (**Figure 10** and **Table 1**) [17]. By using a proper foaming agent the density of the foam was reduced to 0.219–0.458 g/cm^3. The original composition was patented [17].

Depending on the amount of foaming gas appeared, the process involves, in its first phases, only the rupture of the continuous matrix in which the dispersed gas phase is placed (**Figure 11**) [94].

9. Conclusions

1. The structuring of multiphase polymeric materials is the process of arranging the material components in an order in which, the properties of interest are achieved as far as possible. The structuring is achieved by controlling the interface properties so that to be possible the physical-chemical-biological-rheological compatibility of the components considering mainly the melt compounding.

2. To guarantee functional properties of wide application interest, the starch, which as polymer is unprofitable and limited by certain properties, can be converted into multiphase polymeric systems considering melt compounding technique which offers the possibility to dispose the minority components into the majority polymeric matrix in variants with improved compatibility and desired functional properties.

3. The multiphase polymeric systems based on starch can be structured by any method by which the phenomena of the interface are controlled such as: formulations, melt compounding parameters, the flow in the melted state to avoid phenomena as nozzle swelling, shark skin and wall slip, incorporation of target fillers and nano-fillers, incorporation of gaseous phase, the nature and intensity of the homogenization in the melted state, etc. Some possibilities to structure some starch-PVOH compounds are deepened.

4. Structuring of multiphase systems based on starch by reactive compatibilization is connected with designing of the formulation and choosing of the melt compounding conditions which make possible the chemical bonding of dispersed phases and the base polymeric matrix and, in this way, the increasing of miscibility and the level of functional properties so that to be possible the application in sustainable fields of the new obtained starch-based materials.

Acknowledgements

The work on this paper was supported by the Government of Romania, Ministry of Research and Innovation, Project funded by cohesion funds of the European Union Project: P_40_352, SECVENT 81/2016, "Sequential processes of closing the side streams from bioeconomy and innovative (bio) products resulting from it", Subsidiary project 1480/2019 and project supported the Romanian Ministry of Research and Innovation, CCCDI – UEFISCDI, project number no. 23N/2019.

Conflict of interest

The authors declare no conflict of interest.

Author details

Doina Dimonie[1*], Mircea Filipescu[1], Mihai Dragne[2,3], Alina Mustatea[2,3] and Nicoleta Dragomir[1]

1 National Institute for Research and Development in Chemistry and Petrochemistry-ICECHIM, Bucharest, Romania

2 "Applied Chemistry and Materials Science" Doctoral School, Politehnica University of Bucharest, Gheorghe Polizu, Bucharest, Romania

3 PROMATERIS SA, Bucharest, Romania

*Address all correspondence to: ddimonie@yahoo.com

IntechOpen

© 2021 The Author(s). Licensee IntechOpen. This chapter is distributed under the terms of the Creative Commons Attribution License (http://creativecommons.org/licenses/by/3.0), which permits unrestricted use, distribution, and reproduction in any medium, provided the original work is properly cited.

References

[1] Bos H, Elbersen W, Cadórniga Valiño C, Alexopoulou E. Market Demand for Non-Food Crops. 1998. Available from: http://www.cres.gr/crops/pdf/wp6/Final%20Report%20WP6%20Scenarios%20v%2031-01-2011.pdf

[2] Elbersen HW, van der Zee M, Bos HL. The role of 4F crops in EU27 under contrasting future scenarios—Final report on WP6. November 2010. Available from: http://www.cres.gr/crops/pdf/intranet-wp1/WP1_D3_MARKETS.pdf

[3] Zarski A, Bajer K, Kapuśniak J. Review of the most important methods of improving the processing properties of starch toward non-food applications. Polymers. 2021;**13**:832. DOI: 10.3390/polym13050832

[4] Mark JE. Polymer Data Handbook. New York: Oxford University Press, Inc.; 1999. ISBN: 0195107896 9780195107890

[5] McMurry JE. Organic Chemistry. Pacific Grove: Cole Publishing Co.; 1992

[6] Kim Y-K, Rudd ME. Binary-encounter-dipole model for electron-impact ionization. Physical Review. 1994;**50**(5):3954. DOI: 10.1103/PhysRevA.50.3954

[7] Lee SY. Plastic bacteria? Progress and prospects for polyhydroxyalkanoate production in bacteria. Trends Biotechnology. 1996;**14**:431. DOI: 10.1016/0167-7799(96)10061-5

[8] Colonna P. La chimie verte. Paris: Editions TEC&DOC Lavoisier; 2005. ISBN 10: 2-7430-0834-2, ISBN 13: 978-2-7430-0834-5

[9] Struik LCE. Physical Aging in Amorphous Polymers and Other Materials. New York: Elsevier Scientific Publishing Company; 1978

[10] Wilfer PB, Giridaran G, Jeevahan JJ, Joseph GB, G. Kumar S, Thykattuserry NJ. Effect of starch type on the film properties of native starch based edible films. Materials Today: Proceedings. 2021;**44**(5):3903-3907. DOI: 10.1016/j.matpr.2020.12.1118

[11] Shanks, Robert & Kong, Ing. Thermoplastic Starch. 2012. DOI: 10.5772/36295

[12] Yazid NSM, Abdullah N, Muhammad N, Matias-Peralta HM. Application of starch and starch-based products in food industry. Journal of Science and Technology. 2018;**10**(2):144-174. DOI: 10.30880/jst.2018.10.02.023

[13] Nayak PL. Biodegradable polymers: Opportunities and challenges. Journal of Macromolecular Science. 1999;**39**:481-505. DOI: 10.1081/MC-100101425

[14] Doane WM. Starch: Renewable raw material for the chemical industry. Journal of Coatings Technology. 1978;**50**(636):88

[15] Ojogbo E, Ogunsona EO, Mekonnen TH. Chemical and physical modifications of starch for renewable polymeric materials. Materials Today Sustainability. 2020;**7-8**. DOI: 10.1016/j.mtsust.2019.100028

[16] Boudenne A, Ibos L, Candau Y, Thomas S. Handbook of Multiphase Polymer Systems. John Wiley & Sons Ltd; 2011. ISBN:978-0-470-71420-1

[17] Dimonie D. Multipahase Polymeric Materials—Micro-and Nano-Structurated. Bucharest: Politehnica Press; 2019. ISBN: 978-606-515-893-1

[18] Barzic AI, Ioan S. Multiphase Polymer Systems. In: Micro- to Nanostructural Evolution in Advanced Technologies. CRC Press; 2019. ISBN: 9780367876531

[19] Harrats C, Thomas S, Groeninckx G. Micro- and Nanostructured Multiphase Polymer Blend Systems. CRC Press; 2005. ISBN: 0-8493-3734-8

[20] Work WJ, Horie K, Hess M, Stepto RFT. Definitions of terms related to polymer blends, composites, and multiphase polymeric materials (IUPAC recommendations). Pure and Applied Chemistry. 2004;**76**(11):1985-2007. DOI: 10.1351/pac200476111985

[21] Utraki LA. Polymers alloys and blends. State of art. Polymer Networks and Blends. 1991;**1**(2):61-69

[22] Utraki LA. Polymer blends technology. International Journal of Petroleum Science and Technology. 1994;**10**:125-133

[23] Fernandez ML. Demixing in polymer blends. Scientific Progress Oxford. 1990;**74**:257-277

[24] Stroble G. The Physics of Polymers: Concepts for Understanding their Structures and Behavior. Berlin, Heidelberg: Springer; 1996

[25] Hatsua I. Miscibility. In: Encyclopedia of Polymer Science and Engineering. John Wiley &Sons Inc.; 1989

[26] Boudenne A. Handbook of Multiphase Polymer Systems. Wiley; 2012. ISBN: 978-0-470-71420-1

[27] Tager AA, Blinov VS. Thermodynamic compatibility of polymers. Russian Chemical Reviews. 1987;**56**:579

[28] Paul DR. Polymer blends. Journal of Macromolecular Science Reviewers in Macromolecular Chemistry. 1980;**18**(1):109-168

[29] Garton A. Infrared Spectroscopy of Polymer Blends Composites and Surfaces. Carl Hanser Verlag, Munich Publishers; 1992. ISBN: 3-446-17136-3

[30] Malwitz MM, Lin-Gibson S, Hobbie EK, Butler PD, Schmidt GJ. Orientation of platelets in multilayered nanocomposites polymer films. Polymer Science part B Polymer Physics. 2003;**41**:3237

[31] Dimonie D, Kelnar I, Socoteanu R, Darie RN, Pop FS, Zaharia C. The influence of miscibility and micro-structure on the surface defects of some starch bio-hybrides. Materiale Plastice. 2010;**47**(4):486-491

[32] Taguet A, Cassagnau P, Lopez-Cuesta JM. Structuration, selective dispersion and compatibilizing effect of nanofillers in polymer blends. Progress in Polymer Science. 2014;**39**:1526-1563. DOI: 10.1016/j.progpolymsci.2014.04.002

[33] Prentice P. Rheology and its role in plastics processing. Rapra. 1995;7(12):25

[34] Ralph H. Colby, melt viscosity-molecular weight relationship for linear polymers. Macromolecules. 1987;**20**:2226-2237. DOI: 10.1021/ma00175a030

[35] Takeo A, Qui T-C, Mitsubishi S. Structure and Properties of Multiphase Polymeric Materials. CRC Press; 1998. ISBN: 0-8247-0142-9

[36] Wu C-S, Liao H-T. Influence of a compatibilizer on the properties of polyethylene–octene elastomer/starch blends. Journal of Applied Polymer Science. 2002;**86**:1792-1798. DOI: 10.1002/app.11199

[37] Russell CA. Effect of stearate processing aids upon polypropylene stability. Polymer Engineering & Science. 1965;**5**(2):84-89. DOI: 10.1002/pen.760050204

[38] Gachter R, Muller H. Plastics Additives Handbook: Stabilizers, Processing Aids, Plasticizers, Fillers, Reinforcements, Colorants for Thermoplastics. Vol. 754. United States of America: Oxford University Press; 1987. ISBN: 978-3446156807

[39] Sperling LH, Mishra V. Polymeric Multicomponent Materials. New York: Wiley-Interscience Publication, John Wiley & Sons Inc; 1997. pp. 1-25

[40] Mekonnen T, Mussone P. Progress in bio-based plastics and plasticizing modifications. Journal of Materials Chemistry A. 2013;**1**:13379-13398. DOI: 10.1039/C3TA12555F

[41] Dimonie D, Petrache M, Damian C, Anton L, Musat M, Dima Ş-O, et al. New evidences on the process sensitivity of some renewable blends based on starch considering their melt rheological properties. International Journal of Polymer Science. 2016:1-10. DOI: 10.1155/2016/7873120

[42] Dimonie D, Musat M, Doncea SM, Damian CM, Anton L, Vasile E, et al. Controlling the melt resistance to flow as a possibility of improving the miscibility and the time behavior of some blends based on starch. International Journal of Polymer Science. 2015:1-12. DOI: 10.1155/2015/582901

[43] Diyana ZN, Jumaidin R, Selamat MZ, Ghazali I, Julmohammad N, Huda N, et al. Physical properties of thermoplastic starch derived from natural resources and its blends: A review. Polymer. 2021;**13**(9):1396. DOI: 10.3390/polym13091396

[44] Mironescu M, Lazea-Stoyanova A, Barbinta-Patrascu ME, Virchea L-I, Rexhepi D, Mathe E, et al. Green design of novel starch-based packaging materials sustaining human and environmental health. Polymers. 2021;**13**(8):1190. DOI: 10.3390/polym13081190

[45] Encalada K, Aldás MB, Proaño E, Valle V. An overview of starch-based biopolymers and their biodegradability. Revista Ciencia e Ingeniería. 2018;**39**(3):245-258

[46] Bangar SP, Whiteside WS. Nano-cellulose reinforced starch bio composite films- A review ongreen composites. International Journal of Biological Macromolecules. 2021;**185**:849-860. DOI: 10.1016/j.ijbiomac.2021.07.017

[47] Bulatović VO, Mandić V, Grgić DK, Ivančić A. Biodegradable polymer blends based on thermoplastic starch. Journal of Polymers and the Environment. 2021;**29**:492-508. DOI: 10.1007/s10924-020-01874-w

[48] Dammak M, Fourati Y, Tarrés Q, Delgado-Aguilar M, Mutjé P, Boufi S. Blends of PBAT with plasticized starch for packaging applications: Mechanical properties, rheological behaviour and biodegradability. Industrial Crops and Products. 2020;**144**. DOI: 10.1016/j.indcrop.2019.112061

[49] Tai NL, Ghasemlou M, Adhikari R, Adhikari B. Starch-based isocyanate- and non-isocyanate polyurethane hybrids: A review on synthesis, performance and

biodegradation. Carbohydrate Polymers. 2021;**265**:118029. DOI: 10.1016/j.carbpol.2021.118029

[50] Lopez-Gil A, Rodriguez-Perez MA, De Saja JA. Strategies to improve the mechanical properties of starch-based materials: Plasticization and natural fibers reinforcement. Polimeros. 2014;**24**:36-42. DOI: 10.4322/polimeros.2014.053

[51] Jiang T, Duan O, Zhu J, Liu H, Yu L. Starch-based biodegradable materials: Challenges and opportunities. Advanced Industrial and Engineering Polymer Research. 2020;**3**(1):8-18. DOI: 10.1016/j.aiepr.2019.11.003

[52] Xie F, Halley PJ, Avérous L. Rheology to understand and optimize processibility, structures and properties of starch polymeric materials. Progress in Polymer Science. 2012;**37**(4):595-623. DOI: 10.1016/j.progpolymsci.2011.07.002

[53] Pérez MA, Rivas BL, Rodríguez-Llamazares S. Polypropylene/starch blends. Study of thermal and morphological properties. Journal of the Chilean Chemical Society. 2013;**58**:1643-1646. DOI: 10.4067/S0717-97072013000100030

[54] Xie F, Liu P, Yu L. Chapter 10—Processing of Plasticized Starch-Based Materials: State of the Art and Perspectives. In: Halley PJ, Avérous L, editors. Starch Polymers: From Genetic Engineering to Green Application. Elsevier; 2014. pp. 257-289. DOI: 10.1016/B978-0-444-53730-0.00024-5

[55] Mahieu A, Terrié C, Agoulon A, Leblanc N, Youssef B. Thermoplastic starch and poly(ε-caprolactone) blends: Morphology and mechanical properties as a function of relative humidity. Journal of Polymer Research. 2013;**20**:229. DOI: 10.1007/s10965-013-0229-y

[56] Pushpadass HA, Bhandari P, Hanna MA. Effects of LDPE and glycerol contents and compounding on the microstructure and properties of starch composite films. Carbohydrate Polymers. 2010;**82**(4):1082-1089. DOI: 10.1016/j.carbpol.2010.06.032

[57] Matzinos P, Bikiaris D, Kokkou S, Panayiotou C. Processing and characterization of LDPE/starch products. Journal of Applied Polymer Science. 2001;**79**(14):2548-2557. DOI: 10.1002/1097-4628(20010401)79:14<2548::AID-APP1064>3.0.CO;2-3

[58] Van Soest JJG, Vliegenthart JFG. Crystallinity in starch plastics: Consequences for material properties. Trends in Biotechnology. 1997;**15**(6):208-213. DOI: 10.1016/S0167-7799(97)01021-4

[59] Burmistrov VA, Lipatova IM, Losev NV, Rodicheva JA, Koifman OI. Influence of the composition and high shear stresses on the structure and properties of hybrid materials based on starch and synthetic copolymer. Carbohydrate Polymers. 2018;**196**:368-375. DOI: 10.1016/j.carbpol.2018.05.056

[60] Liu H, Xie F, Yu L, Chen L, Li L. Thermal processing of starch-based polymers. Progress in Polymer Science. 2009;**34**(12):1348-1368. DOI: 10.1016/j.progpolymsci.2009.07.001

[61] Liu Y, Chen X, Xu Y, Xu Z, Li H, Sui Z, et al. Gel texture and rheological properties of normal amylose and waxy potato starch blends with rice starches differing in amylose content. International Journal of Food Science and Technology. 2021;**56**:1946-1958. DOI: 10.1111/ijfs.14826

[62] Giles HF, Wagner JR, Mount EM. Extrusion: The Definitive Processing Guide and Handbook. Plastics Design Library. Norwich, NY: William Andrew Pub; 2005

[63] Coserea RM, Dimonie D, Dimonie M, Hubca G. Some aspects concerning the rheology of biodegradable starch based materials. University Politehnica of Bucharest Scientific Bulletin Series B-Chemistry and Materials Science. 2012;**74**(2):37-48

[64] Dimonie D, Damian C, Trusca R, Rapa M. Some aspects conditioning the achieving of filaments for 3D printing from physical modified corn starch. Materiale Plastice. 2019;**56**(2):351-359

[65] Dimonie D, Petre D, Vasilievici G. Polyvinyl alcohol melt processing. Journal of Elastomers and Plastics. 2007;**39**:181-194. DOI: 10.1177/0095244306071978

[66] Fourati Y, Tarrés Q, Mutjé P, Boufi S. PBAT/thermoplastic starch blends: Effect of compatibilizers on the rheological, mechanical, and morphological properties. Carbohydrate Polymers. 2018;**199**:51-57. DOI: 10.1016/j.carbpol.2018.07.008

[67] Mochane MJ, Sefadi JS, Motsoeneng TS, Mokoena TE, Mofokeng TG, Mokhena TC. The effect of filler localization on the properties of biopolymer blends, recent advances: A review. Polymer Composites. 2020;**41**:2958-2979. DOI: 10.1002/pc.25590

[68] Maliger RB, McGlashan SA, Halley PJ, Matthew LG. Compatibilization of starch–polyester blends using reactive extrusion. Polymer Engineering and Science. 2006;**46**(3):248-263. DOI: 10.1002/pen.20479

[69] Schwach E, Avérous L. Starch-based biodegradable blends: morphology and interface properties. Polymer International. 2004;**53**(12):2115-2124. DOI: 10.1002/pi.1636

[70] Davoodi S, Oliaei E, Davachi SM, Hejazi I, Seyfi J, Heidari BS, et al. Preparation and characterization of interface-modified PLA/starch/PCL ternary blends using PLLA/triclosan antibacterial nanoparticles for medical applications. RSC Advances. 2016;**6**:39870-39882. DOI: 10.1039/C6RA07667J

[71] Imre B, García L, Puglia D, Vilaplana F. Reactive compatibilization of plant polysaccharides and biobased polymers: Review on current strategies, expectations and reality. Carbohydrate Polymers. 2019;**209**:20-37. DOI: 10.1016/j.carbpol.2018.12.082

[72] Hu H, Xu A, Zhang D, Zhou W, Peng S, Zhao X. High-toughness poly (lactic acid)/starch blends prepared through reactive blending plasticization and compatibilization. Molecules. 2020;**25**(24):5951. DOI: 10.3390/molecules25245951

[73] Shin BY, Narayan R, Lee SI, Lee TJ. Morphology and rheological properties of blends of chemically modified thermoplastic starch and polycaprolactone. Polymer Engineering and Science. 2008;**48**(11):2126-2133. DOI: 10.1002/pen.21123

[74] Marinho VAD, Pereira CAB, Vitorino MBC, Silva AS, Carvalho LH, Canedo EL. Degradation and recovery in poly(butylene adipate-co-terephthalate)/thermoplastic starch blends. Polymer Testing. 2017;**58**:166-172. DOI: 10.1016/j.polymertesting.2016.12.028

[75] Hwang SW, Shim JK, Selke S, Soto-Valdez H, Rubino M, Auras R.

Effect of maleic-anhydride grafting on the physical and mechanical properties of poly (L-lactic acid)/starch blends. Macromolecular Materials and Engineering. 2013;**298**:624-633. DOI: 10.1002/mame.201200111

[76] Jang WY, Shin BY, Lee TJ, Narayan R. Thermal properties and morphology of biodegradable PLA/starch compatibilized blends. Journal of Industrial and Engineering Chemistry. 2007;**13**(3): 457-464

[77] Lendvai L, Brenn D. Mechanical, morphological and thermal characterization of compatibilized poly (lactic acid)/thermoplastic starch blends. Acta Technica Jaurinensis. 2020;**13**(1):1-13. DOI: 10.14513/actatechjaur.v13.n1532

[78] Bher A, Uysal Unalan I, Auras R, Rubino M, Schvezov C. Toughening of poly(lactic acid) and thermoplastic cassava starch reactive blends using graphene nanoplatelets. Polymers. 2018;**10**:95. DOI: 10.3390/polym10010095

[79] Madhumitha G, Fowsiya J, Mohana Roopan S, Thakur VK. Recent advances in starch–clay nanocomposites. International Journal of Polymer Analysis and Characterization. 2018;**23**(4):331-345. DOI: 10.1080/1023666X.2018.1447260

[80] Avella M, Vlieger JJ, Errico ME, Fischer S, Vacca P, Volpe MG. Biodegradable starch/clay nanocomposite films for food packaging applications. Food Chemistry. 2005;**93**(3):467-474. DOI: 10.1016/j.foodchem.2004.10.024. AGR: IND43721001

[81] García NL, Ribba L, Dufresne A, Aranguren M, Goyanes S. Effect of glycerol on the morphology of nanocomposites made from thermoplastic starch and starch nanocrystals. Carbohydrate Polymers. 2011;**84**(1):203-210. DOI: 10.1016/j.carbpol.2010.11.024

[82] Xie F, Pollet E, Halley PJ, Avérous L. Advanced Nano-Biocomposites Based on Starch. In: Ramawat K, Mérillon JM, editors. Polysaccharides. Cham: Springer; 2014

[83] Hietala M, Mathew AP, Oksman K. Bionanocomposites of thermoplastic starch and cellulose nanofibers manufactured using twin-screw extrusion. European Polymer Journal. 2013;**49**(4):950-956. DOI: 10.1016/j.eurpolymj.2012.10.016

[84] Boonprasith P, Wootthikanokkhan J, Nimitsiriwat N. Mechanical, thermal, and barrier properties of nanocomposites based on poly(butylene succinate)/thermoplastic starch blends containing different types of clay. Journal of Applied Polymer Science. 2013;**130**:1114-1123. DOI: 10.1002/app.39281

[85] García NL, Ribba L, Dufresne A, Aranguren MI, Goyanes S. Physico-mechanical properties of biodegradable starch nanocomposites. Macromolecular Materials and Engineering. 2009;**294**:169-177. DOI: 10.1002/mame.200800271

[86] McGlashan SA, Halley PJ. Preparation and characterisation of biodegradable starch-based nanocomposite materials. Polymer International. 2003;**52**:1767-1773. DOI: 10.1002/pi.1287

[87] Vasile C, Stoleriu A, Popescu M-C, Duncianu C, Kelnar I, Dimonie D. Morphology and thermal properties of some green starch/poly (vinyl alcohol) / montmorillonite nanocomposites. Cellulose Chemistry and Technology. 2008;**42**(9-10):549-568

[88] Dimonie D, Radovici C, Vasilievici G, Popescu MC, Garea S. The dependence of the XRD Morphology of some bionanocomposites on the silicate treatment. Journal of Nanomaterials. 2008;**2008**:1687-4110. DOI: 10.1155/2008/538421

[89] Dimonie D, Radovici C, Zaharia C, Vasilievici G, Stoleriu A. Comportarea termica a nanocompozitelor biodegradabile pe baza de alcool polivinilic si amidon. Materiale Plastice. 2006;**43**(2):132-137

[90] Dimonie D, Radovici C, Trandafir I, Pop FS, Dumitru I, Fierascu R, et al. Some aspects concerning the silicat delamination for obtaining polymeric bio-hybrids based on starch. Revue Roumaine de Chimie. 2011;**56**(7):685-690

[91] Dimonie D, Radovici C, Pop FS, Socoteanu R, Petre C, Fierascu R, et al. Approaches considering a better incorporations of multylayers silicate into a polymeric matrix designed for bio–nano–materials obtaining. Nonlinear Optics, Quantic Optics: Concept in Modern Optics. 2012;**44**(2-3):137-148

[92] Dimonie D, Socoteanu R, Doncea SM, Pop FS, Petrea C, Fierascu R, et al. Miscibility estimation of some blends based on starch. E-Polymers. 2011;**11**(1):957-970. DOI: 10.1515/epoly.2011.11.1.957

[93] Dimonie D, Kelnar I, Duncianu C, Coserea RM, Pop FS, Dumitriu I, et al. Some aspects concerning the surface defects of green nanocomposites based on renewable resources. In: Proceeding of 16th Romanian International Conference on Chemistry and Chemical Engineering (16th RICCCE); September 9-12. Sinaia: Ed. Printech; 2009. ISBN 978-606-521-349-4

[94] Dimonie D. RO Compozite biodegradabile, pentru ambalaje nealimentare si procedeu de obtinere a acestora. Patent No. 121692/28.02.2008

Chapter 11

Starch-Based Hybrid Nanomaterials for Environmental Remediation

Ashoka Gamage, Thiviya Punniamoorthy and Terrence Madhujith

Abstract

Environmental pollution is becoming a major global issue with increasing anthropogenic activities that release massive toxic pollutants into the land, air, and water. Nanomaterials have gained the most popularity in the last decades over conventional methods because of their high surface area to volume ratio and higher reactivity. Nanomaterials including metal, metal oxide, zero-valent ions, carbonaceous nanomaterials, and polymers function as adsorbents, catalysts, photocatalysts, membrane (filtration), disinfectants, and sensors in the detection and removal of various pollutants such as heavy metals, organic pollutants, dyes, industrial effluents, and pathogenic microbial. Polymer-inorganic hybrid materials or nanocomposites are highly studied for the removal of various contaminants. Starch, a heteropolysaccharide, is a natural biopolymer generally incorporated with other metal, metal oxide, and other polymeric nanoparticles and has been reported in various environmental remediation applications as a low-cost alternative for petroleum-based polymers. Therefore, this chapter mainly highlights the various nanomaterials used in environmental remediation, starch-based hybrid nanomaterials, and their application and limitations.

Keywords: environmental remediation, hybrid nanomaterials, nanomaterials, starch, starch-based hybrid nanomaterials

1. Introduction

Environmental pollution is becoming a serious global problem that society faces today. Ongoing anthropogenic activities, extensive food and agriculture practices, industrialization, and urbanization release huge amounts of pollutants into the environment that can cause air, water, and land pollution, consequently threatening to human, animal health, and ecosystem [1, 2]. These toxic pollutants can enter the human body either through inhalation, ingestion, or absorption and adversely affect health. Further, bioaccumulation of some heavy metals through the food chain and persistent organic pollutants in biota and fishes poses a huge threat to humans and wildlife and requires sustainable, efficient, and low-cost technologies to detect, monitor, and remediate the hazardous pollutants [1].

Different forms of pollutants are released into the environment; soil, water, and air. Organic substances (pesticides, insecticides, fertilizers, oil spills, phenols, chloroform, hydrocarbons), heavy metals and metalloids (Cr^{2+}, Pb^{2+}, Co^{2+}, Cd^{2+}, Cu^{2+}, Zn^{2+}, Mn^{2+}, Ni^{2+}, As, Hg), dyes, industrial effluents, sewage, as well as microbial pathogens are few contaminants in soil and water. While, contaminants such as toxic gases (nitrogen oxides, sulfur oxides, carbon oxides, ozone), suspended airborne particles, and volatile organic compounds are found in the atmosphere [3, 4].

These contaminants in soil, water, and air are remediated by using different conventional techniques, such as physical, chemical, and biological methods [4–6]. These techniques may be used in combination with one another to remediate contaminated sites. Adsorption (clay minerals, industrial wastes, biomass, biochar, activated carbon, biopolymer), chemical treatments, bioremediation, coagulation and flocculation, ion exchange, membrane-filtration, solidification/stabilization, electrokinetics, and electrochemical treatments technologies have been used in heavy metal removal from soil and water [7]. Bioremediation using microorganisms and plants helps to detoxify or remove crude oil, heavy metal removal, and pesticide degradation from soil and water [4, 5].

However, the majority of these conventional techniques are expensive, laborious, environmentally destructive, time-consuming methods, also involved in the consumption of chemicals and the generation of undesirable toxic by-products that are hazardous to the environment. Further, complexities of the mixture of different compounds, high volatility, and low reactivity of contaminants also limit the applications in environmental remediation [3, 5, 8]. New environmental remediation technologies are constantly being explored, and recent studies have focused on developing new environmental remediation technologies using various nanomaterials [3].

2. Nanotechnology in environmental remediation

2.1 Nanotechnology and its advantages and applications

Nanotechnology has gained much attention in environmental remediation over the last few decades [1]. Nanotechnology is an advanced technology that works on the material in nanometer scale (1–100 nm) and produces materials, devices, and systems with specific and novel properties and functions by controlling the size and the shape of matters [1, 4, 9]. The nanomaterials are broadly categorized as organic and inorganic nanomaterials. Some literatures is classified based on materials used in the synthesis process; inorganic (metal, metal oxide, zero-valent metals), carbon-based [graphene, carbon nanotubes (CNTs)], polymer-based (dendrimers or polyamidoamine), and composite based nanomaterials [3, 10].

Nanomaterials have several advantages in environmental remediation over conventional methods; cost-effective, simple to use, energy conservative, sustainable, and more effective methods. Due to the properties such as smaller size (1–100 nm) and higher surface area to volume ratio of nanomaterials, they provide more reaction surface area, which increases reactivity and thus its sensitivity and effectiveness. Nanoparticles have a high sorption capacity for inorganic and organic compounds because of their specific characteristics; large surface area, an increased number of surface activation sites, a good affinity to other species [11]. Further, nanotechnology helps in the development of remediation technologies that are specific and efficient for a particular pollutant [3, 9].

Nanotechnology has potential applications in many fields, including food and agriculture, packaging, pharmaceutical, drug delivery, energy, and pollution treatment [1, 12]. Of which, the application of nanotechnology in pollution control and environmental remediation has gained popularity over the last decade; wastewater treatment, cleaning groundwater, and remediation of soil contaminated with pollutants. In the field of environment, nanotechnology has been used in pollution detection (sensing and detection), prevention of pollution, and purification/remediation of contamination [9]. Thus, nanotechnology provides a sustainable solution to the global challenges of protecting water, soil and providing cleaner air [13].

2.2 Nanomaterials in environmental remediation

Various nanomaterials such as inorganic, carbonaceous nanomaterials, polymer-based nanomaterial are used in environmental remediation (air, soil, and water) as adsorbents, catalyst, photocatalyst, membrane (filtration), disinfectants, and sensors [1, 3, 14].

Metal (silver, gold), metal oxides (iron oxides, TiO_2, MgO, Fe_2O_3, Al_2O_3), and zero-valent metals (Fe^0, Zn^0, Sn^0, and Al^0) based nanoparticles are mostly studied for environmental remediation including disinfection of water, treatment of drinking water, groundwater, wastewater, and air, because of their adsorption, antibacterial, antimicrobial, photocatalytic, reductive dehalogenation, desulfurization, and catalytic reduction activities [3, 15, 16]. Carbonaceous materials in different structural configurations; fullerene, single-walled carbon nanotubes (SWCNTs), multi-walled carbon nanotubes (MWCNTs), and graphene and used in the removal of organic and inorganic contaminants from air and water due to its adsorption and photocatalytic property [3].

Nanoscale zero-valent iron (nZVI) is the most widely studied nanoparticle in soil remediation [12] and is used for reductive immobilization of heavy metals in soil that decreases the bioavailability and mobility of heavy metals and prevents leaching into groundwater and transfers to the food chain [1]. Further, nanomaterials such as nanoparticles (NPs) (metal; Au, Ag, Fe, bimetal; Fe/Ni, Ag/Cu, metal oxides; TiO_2, ZnO, Fe_2O_3), nanotubes (carbon nanotubes, halloysite nanotubes), and nanocomposites (graphene oxide) have been reported to utilize in detection, degradation, and removal by adsorption of pesticides [17].

Emission of greenhouse gases (carbon dioxide, methane, nitrous oxide, and fluorinated gases), volatile organic compounds (ethylene, aniline, benzene), are controlled either by separation or capturing, such as filtration, absorption in liquids, adsorption on solids, or a combination of these processes. In addition, bioaerosols (aerosols of biological origin such as viruses, bacteria, and fungi), an indoor air pollutant, can rapidly spread with airflow and can cause numerous diseases, including infections and allergies. The air filtration process using antimicrobial materials such as Ag NPs, Cu NPs, CNTs, and natural products is the most applied and effective technique to remove bioaerosols [1].

Various nanomaterials have also been studied for the treatment of drinking water and industrial wastewater, including adsorbents (nZVI or Fe, MnO, ZnO, MgO, Al_2O_3, TiO_2, Magnetite or Fe_3O_4, CNT), photocatalysts (ZnO, TiO_2, metal-based nanocomposites such as Ag/ZnO and Pt/ZnO, CdS, ZnS: Cu, CdS: Eu, CdS: Mn), electrocatalysts (Pt, Pd, Au/metal oxides TiO_2, MgO, Fe_2O_3, Al_2O_3), nano-membranes (MWCNTs, electrospun PVDF, PVC, sodium titanate nanobelt membrane), disinfectants with antibacterial effects (Ag NPs, chitosan NPs, TiO_2), nanosensors (Au NPs, Ag NPs) [14, 16, 18].

2.3 Hybrid nanomaterials

The term hybrid refers to fusion, joining, or mixing of characteristics at the molecular level, which generates a hybrid material owning the effective functionality of single components and eliminates undesirable characteristics [19, 20]. In this context, hybrid nanomaterials are defined as materials that are made up of two or more organic or inorganic components such as organic-organic (starch-cellulose), inorganic-inorganic (TiO_2-Ag), and organic-inorganic (starch-TiO_2) compounds, connected at the nanometer scale, combine the intrinsic characteristics of its individual constituents to additional properties due to synergistic effects between the components [21, 22]. These hybrid materials are synthesized by different methods such as covalent immobilization, electrostatic binding, polymerization methods, among others [21]. The properties of the hybrid material vary with the material (organic or inorganic), structure, and different component interface, and the optimum combination can enhance mechanical strength and thermosensitivity, improve thermal and chemical stability, and regulate optical, anticorrosive, magnetic, electrical, and thermal properties as well as fire retardancy [23]. Because of their excellent mechanical, physical, and tribological characteristics, hybrid nanomaterials are widely used in the area of food packaging, plant protection, electrochemistry, and various additional applications in the environmental, biotechnological, and agri-food sectors [19].

Generally, hybrid materials are classified into two categories depending on the intra- and intermolecular interactions among the organic matrix and cross-linking agent [21, 23];

1. Class I (organic and inorganic exhibiting weaker interactions such as noncovalent interactions; van der Waals and hydrogen bonding).

2. Class II (organic and inorganic exhibiting strong interactions such as covalent, ionic, ionocovalent, and coordinative bonding).

"Polymer-based composites" or "nanocomposites" can be defined as hybrid organic-inorganic composites when incorporating either component in nanoscale and generally obtained by incorporation of a small quantity of an inorganic component into an organic or a polymer matrix in order to form a new component with enhanced properties [24]. The "bio-nano composites" are the materials that comprise particles with at least one dimension in the range of 1–100 nm and a constituent(s) of the biological origin or maybe biopolymers.

Biopolymers (natural polymers) have received much attention in recent last decades due to their abundance, low toxicity, low cost, biodegradability, biocompatibility, and multiple functionalities [25]. A variety of biopolymers such as polysaccharides (cellulose, chitin, chitosan, pectin, starch, dextran, xanthan, guar gum, fucoidan, heparin, hyaluronan, and pullulan), proteins (albumin, casein, collagen, fibrinogen, and gelatin), polylactic acid (PLA), and nucleic acids have been used as alternative eco-friendly materials to replace synthetic polymers or petroleum-based polymers (PP, PE, and epoxies) partially or even totally [25–27]. Polysaccharide-based hybrid nanocomposites have become increasingly essential materials over the past decades [25, 27]. Many studies have reported the application of polysaccharide-based nanocomposites (natural polymer) in various fields such as food, biomedical, ecofriendly and sustainable food packaging, and environmental pollution control and remediation [28–30].

Due to the poor barrier, mechanical, and processing properties, natural polymers (biopolymers) are incorporated with other synthetic polymers or nanomaterials to improve their properties and applications [31]. Polysaccharides such as cellulose, chitin, chitosan, and starch are the most studied biopolymers and used in biodegradable nanocomposites with metal nanoparticles (Au, Ag, Cu, and Pd), metal oxide nanoparticles (TiO_2, ZnO, CuO, Cu_2O, SiO_2, Fe_2O_3, and Fe_3O_4) and carbon nanomaterials (graphene and carbon nanotubes, CNTs) [25].

3. Starch hybrid nanomaterials for environmental remediation

3.1 Starch

Starch, a natural, abundant, renewable, biocompatible, and biodegradable biopolymer, is naturally found in many plants as the primary source of energy and reserved in many parts of plants such as stalks, stems, roots, tubers, and seeds; main sources being cassava, wheat, rice, barley, maize or corn, banana, and potatoes, among others. Starch is a heteropolysaccharide that comprises D-glucose monomers joined with glycosidic bonds and can be denoted as $(C_6H_{10}O_5)_n$ with the basic chemical formula. Starch is a heteropolysaccharide composed of two types of macromolecules: linear amylase (around 10–30% of starch granule) and branched amylopectin (remaining 70–90% of starch granule). Amylose is a linear polysaccharide chain of D-glucose units linked by α-(1,4)-glycosidic bond with a degree of polymerization in a range of 300–10,000. Amylopectin is a very high-molecular-weight polymer with a backbone structure of amylase cross-linked through α-(1,6) glycosidic bonds. The basic structure of amylose and amylopectin are shown in **Figure 1** [25, 32, 33].

3.2 Starch hybrid nanomaterials

Starch-based nanocomposites have wide applications in the fields of food and agriculture, packaging, biomedical, and environmental remediation as emulsion stabilizers, fat replacers, flexible films, carriers of bioactive compounds, drug delivery, and adsorbents in sewage treatment or wastewater treatment [34–36]. Starch nanoparticles are usually smaller than 300 nm in dimension with a high specific surface area. The various forms of starch-based nanoparticles are starch nanoparticles, starch nanospheres, starch micelles, starch vesicles, starch nanogels, and starch nanofibers [36].

Starch is a natural polymer, gained much attention because of its renewability, biodegradability, abundance, eco-friendly, relatively low cost, non-toxic, high adsorptive capacities, amenable to various chemical modifications, and cohesive film-forming properties. Starch molecules can bind with the heavy metal ions or contaminants through the functional (hydroxyl) groups on the starch structure [37, 38]. Further, high amylopectin content in starch has powerful swelling properties that are important in sorption-based applications [39]. In most published works, carbohydrates have been used as reducing, stabilizing, and/or complexing agents [40].

However, starch in a pure or native form has drawbacks such as poor processability, high brittleness, susceptibility to retrogradation, high viscosity, low adsorption capacity, and greater hydrophilicity or high-water absorption capacity, which limits its many applications in the environmental field. To overcome this problem and to obtain water-insoluble materials, starch is modified by physically [hydrothermal

(a)

(b)

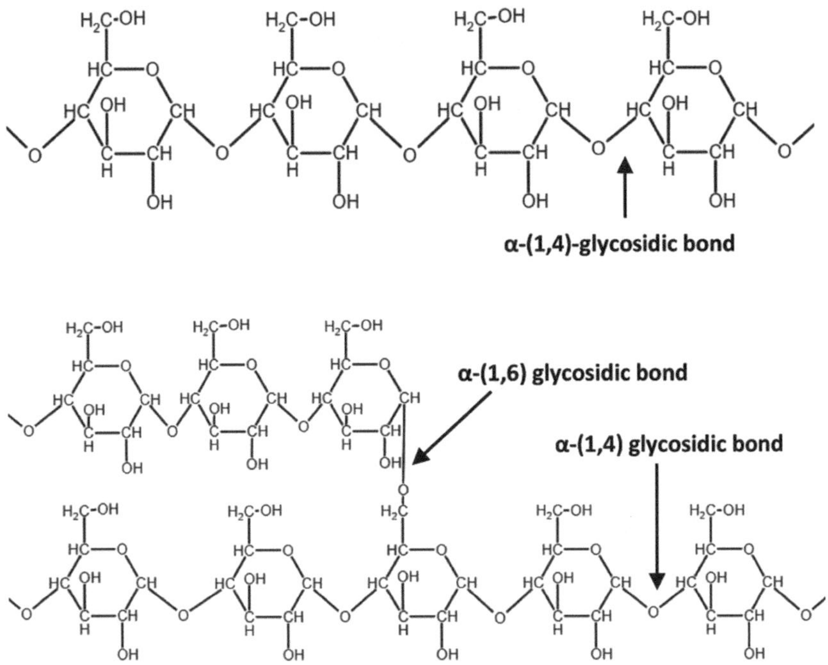

Figure 1.
Structures of starch: (a) amylose and (b) amylose pectin.

processing (i.e. gelatinization)] or chemically (etherification, esterification, cross-linking, grafting, oxidation, and enzymatic hydrolysis) or a combination of these two methods [41–44]. Polysaccharides exhibit a great number of reactive hydroxyl groups, which can be exploited for direct esterification, etherification, and various chemical modifications [41].

Starch-based hybrid materials have numerous functionalities and/or novel properties due to the interactions between the individual constituents, mostly associated with synergetic effects, and have been reported in environmental remediation applications [25]. Several starch-based composites have been reported to have a remarkable adsorption tendency for the removal of heavy metals and dyes [45].

3.3 Starch-based hybrid nanomaterials in environmental remediation

3.3.1 Starch/metal or metal oxides or non-valent metals

Table 1 shows the recent examples of the combination of starch and different metal, metal oxide, zero-valent metal, CNTs, and other polymers nanoparticles, such as Au, Ag, Cu, Pd, ZnO, TiO_2, nZVI, among others. Nanomaterials are widely used to treat different contamination because of their high specific surface area to volume ratio, rapid kinetics, and high reactivity. However, pure or unmodified nanoparticles tend to agglomerate easily into larger particles that decrease the available specific surface area and reactivity. To improve the colloidal stability of nanoparticles, surface modification has been done by coating with various polymers. Of which starch is one of the relatively cheap and green polysaccharides [53, 59].

Starch hybrid nanomaterials	Application	Reference
Starch-stabilized Fe/Cu	As^{2+} and As^{5+} removal from the contaminated water	[46]
Starch/Fe_3O_4	Removal of perfluorooctanoic acid (PFOA) in soil and groundwater	[47]
Starch/nZVI	Soil remediation—heavy metal removal, particularly Pb and Zn	[48]
Starch/nZVI	Removal of As^{3+} and As^{5+} from aqueous solutions	[49]
Starch-modified nZVI	Removal of Cr^{6+}	[50]
Starch/SnO_2	Water treatment—Hg^{2+}	[51]
Starch/TiO_2	Removal and determination of heavy metals such as Cd, Co, Cu, Pb, and Ni	[11]
Starch/glycidyl methacrylate	Adsorption of Pb^{2+}, Cd^{2+}, Cu^{2+} and Cr^{3+} in wastewater	[52]
Modified tapioca starch coated nZVI	Aqueous nitrate remediation	[53]
Starch-Fe-Pd	Water decontamination—trichloroethene	[37]
Starch xanthate	Water decontamination—Ni^{2+}, Cu^{2+}, Cr^{3+}	
Carboxymethyl starch	Water decontamination—Pb^{2+}, Cd^{2+}, Cu^{2+}	
Polymerized starch with epichlorohydrin	Water decontamination—dyes	
Starch/polyvinyl alcohol (PVA)	Textile wastewater treatment—dye including methylene blue (cationic dye), methylene orange (anionic dye), starch, PVA	[54, 55]
PVA/corn starch hydrogel	Iron (Fe^{3+}) and arsenic (As^{3+}) removal in aqueous solution	[56]
Starch/TiO_2	Removal of dye from water/wastewater	[32]
Starch-NiFe-layered double hydroxide	Adsorption of anionic dye methyl orange (MO) from aqueous solution	[45]
Starch/poly(alginic acid-cl-acrylamide) nanohydrogel	Adsorption of coomassie brilliant blue R-250 dye from the aqueous solution	[57]
Polyaniline/starch/hematite (PANI/starch/Fe_2O_3)	Wastewater remediation—adsorption of different heavy metals, including As^{3+}, Zn^{2+}, and Co^{2+} and antibacterial effect	[38]
MWCNT-starch-iron oxide	Adsorbent for removing methyl orange (MO) and methylene blue (MB) from aqueous solutions	[58]

Table 1.
Various starch-based hybrid nanomaterials and their applications in environmental remediation.

Rashid et al. reported that modified tapioca starch could be used as an effective surface modifier for nZVI particles for aqueous nitrate removal [53]. Starch-stabilized Fe/Cu nanoparticles in arsenic (As^{2+} and As^{5+}) removal from the contaminated water where Cu as a metal catalyst was incorporated with Fe^0 (nZVI) to form an iron bimetallic nanoparticle; then, the surface was modified to prevent the agglomeration [46].

Well stabilized (dispersed) iron oxides nanoparticles offer greater specific surface area and sorption capacity than the nanoparticles without any stabilizer towards a wide range of pollutants. Starch-functionalized magnetite (Fe_3O_4) nanoparticles showed much higher As^{2+} and As^{5+} sorption capacity than pristine magnetite nanoparticles [59]. Starch-stabilized Fe_3O_4 nanoparticles can be used as a "green" adsorbent for the effective removal of perfluorooctanoic acid (PFOA) in soil and groundwater [47]. Baysal et al. reported that starch-coated TiO_2 NPs can be successfully used as adsorbents for the removal and determination of heavy metals such as Cd, Co, Cu, Pb, and Ni [11]. The starch-based SnO_2 nanocomposite material can be used as an adsorbent for the removal of highly toxic Hg^{2+} metal ions from an aqueous medium [51].

3.3.2 Starch/carbon nanotubes (CNTs)

CNTs have gained increased attention in multidisciplinary studies because of their unique physical and chemical properties. However, the hydrophobicity of CNTs may limit their application. The hydrophilicity and biocompatibility of CNTs can be improved by incorporating biopolymers such as starch in the composite system. Incorporating CNTs with starch also helps to overcome the limitation of starch, i.e. weak mechanical properties and poor long-term stability [60, 61]. MWCNT-starch-iron oxide has been reported as a better adsorbent for removing anionic dye methyl orange (MO) and cationic dye methylene blue (MB) from aqueous solutions than MWCNT-iron oxide. The hydrophilic property of soluble starch improved the hydrophilicity of MWCNTs and the dispersion of MWCNT-starch-iron oxide in the aqueous solution. In addition, the increased contact surface between magnetic MWCNT and dyes reduced the aggregates of MWCNTs and facilitated the diffusion of dye molecules to the surface of MWCNTs. Nanoparticles, ZnO, TiO_2, or Ag or their complex decompose the adsorbed organic contaminants on MWCNTs as the photocatalysts [60].

3.3.3 Blending starch nanoparticles with different biopolymeric matrices

Starch-based hydrogels have a good adsorption capacity, which can be used for wastewater treatment by removing various cationic or anionic dyes after modification with functional groups [44]. The incorporation of starch into synthetic polymer hydrogel networks improves their swelling and adsorption capacity [44]. Hydrogel as an adsorbent is one of the best candidates for removing soluble dyes from an aqueous solution. The study of methylene blue (MB) adsorption efficiency of NaOH-treated starch/ acrylic acid hydrogel showed high dye-capturing coefficients, which increase with the starch ratio and indicates the possibility of the hydrogels' application for removing dyes from aqueous solution. In which, starch can be a natural-polymer superabsorbent because of a large number of hydrophilic groups (–OH) and other benefits such as renewable, very cheap, and biodegradable [62]. Biodegradable polymers, starch/cellulose nanowhiskers hydrogel composite, showed outstanding adsorption capacity to be employed in the remediation of methylene blue contaminated wastewaters [63]. Pectin-starch magnetite hybrid nanoparticles could be potential adsorbents for methylene blue dye with higher adsorption efficiency at a low polymer concentration and starch-pectin ratio and can be used to recycle water from the textile industry [58].

3.4 Limitations and future studies for using starch hybrid nanomaterials in environmental remediation

Increased nano-waste release in the environment, bioaccumulation, occupational exposure, and nanotoxicity are the major problems associated with the increased use of nanomaterials in environmental remediation. Nanoparticles incorporated in starch-based hybrid nanomaterials such as Ag, Au, nZVI, TiO_2, SiO_2, ZnO, Al_2O_3, CNTs, metal chalcogenides (CdS, CdSe), polymeric nanoparticles, among others, shows toxicity (acute or chronic) in high dose; growth inhibition of microalgae, disruption of membrane integrity, reactive oxygen species generation, oxidative stress, genotoxicity, and mutagenicity up to reproduction impairment in aquatic species and many health complications in human [41, 64–67].

Because of the very small size, nanoparticles are capable of entering the human body by inhalation, ingestion via food, drink, and drugs, skin penetration, or injections and they have the potential to interact with intracellular structures and macromolecules for long periods [68]. Exposure to nanoparticles is associated with a range of acute and chronic effects ranging from inflammation, exacerbation of asthma, and metal fume fever to fibrosis, chronic inflammatory lung diseases, and carcinogenesis [64].

The effect of surface modification of nanoparticles such as nZVI is not clear. Sun et al. reported that surface modifiers enhance the stability of the nZVI that either increase the toxicity due to prolonged exposure to the living organisms or decrease the toxicity via reducing the adhesion of nZVI to living organisms or preventing the release of toxic ions. Starch stabilized nZVI produced higher phytotoxicity compared to bare nZVI, this may be due to the higher dispersity, hydrophilicity, and anti-aggregation of starch/nZVI that enhances their affinity to root surfaces and the oxidability of the Fe^0, forming a coating of insoluble Fe^{3+} compounds on the root surface, and thus interferes the absorption of water and nutrients [69].

In the future, attention will be given to the green synthesis of nanomaterials because not all nanomaterials are produced in an eco-friendly way, as involves acid hydrolysis in multiple steps. There are several systems and methods for the green synthesis of nanoparticles, particularly enzymes, vitamins, microwave, bio-based methods, and from plants and phytochemicals [67, 70]. Green synthesis of nanoparticles using various natural sources, non-toxic solvents, and techniques (ultrasound, microwave, hydrothermal, magnetic, and bioproduction by fungi and other microorganisms) promote eco-friendly, sustainable, less expensive, and free of chemical contaminant production and applications [68].

Nano-wastes should be diluted and neutralized before disposal as they are extraordinarily toxic, hazardous, and/or chemically reactive. Proactive nano-waste management strategies need to be adopted to prevent long-term unintended consequences, and, where possible, nano-waste should be recycled [64].

4. Conclusion

Remediation is the science of removal or reduction of pollutants from the environment using chemical or biological means. Starch-based hybrid materials are a cost-effective and eco-friendly solution over petroleum-based polymers in environmental remediation. Though starch is a natural polymer with many benefits, including renewability, biodegradability, abundance, eco-friendly, relatively low cost, non-toxic, poor barrier, and mechanical properties, poor processability, high brittleness, and high hydrophilicity are major drawbacks of raw starch. Therefore, starch is modified by physical and/or chemical methods, including gelatinization, etherification, esterification, crosslinking, grafting, oxidation, and enzymatic hydrolysis.

Starch-based hybrid materials have numerous functionalities and/or novel properties, mainly associated with synergetic effects and reported in environmental remediation applications. Starches are incorporated with metal NPs, metal oxide NPs, zero-valet metals, CNTs, and other polymers as reducing, stabilizing, and/or complexing agents to remove various toxic contaminants such as heavy metal, organic contaminants, and dye wastewater and groundwater.

In future studies, various natural starch sources, green synthesis of nanomaterials, recyclability, and toxicity effect of nano-waste should be considered. Further development of biodegradable starch-based hybrids and nanomaterials focusing on new functional materials, processing technology, and cost reduction needs to be studied for commercial application.

Conflict of interest

The authors declare no conflict of interest.

Author details

Ashoka Gamage[1*], Thiviya Punniamoorthy[2] and Terrence Madhujith[3]

1 Faculty of Engineering, Department of Chemical and Process Engineering, University of Peradeniya, Sri Lanka

2 Postgraduate Institute of Agriculture, University of Peradeniya, Sri Lanka

3 Faculty of Agriculture, Department of Food Science and Technology, University of Peradeniya, Sri Lanka

*Address all correspondence to: ashogamage@gmail.com

IntechOpen

© 2021 The Author(s). Licensee IntechOpen. This chapter is distributed under the terms of the Creative Commons Attribution License (http://creativecommons.org/licenses/by/3.0), which permits unrestricted use, distribution, and reproduction in any medium, provided the original work is properly cited.

References

[1] Ibrahim RK, Hayyan M, AlSaadi MA, Hayyan A, Ibrahim S. Environmental application of nanotechnology: Air, soil, and water. Environmental Science and Pollution Research. 2016;**23**:13754-13788. DOI: 10.1007/s11356-016-6457-z

[2] Mózner Z, Tabi A, Csutora M. Modifying the yield factor based on more efficient use of fertilizer—The environmental impacts of intensive and extensive agricultural practices. Ecological Indicators. 2012;**16**:58-66. DOI: 10.1016/j.ecolind.2011.06.034

[3] Guerra FD, Attia MF, Whitehead DC, Alexis F. Nanotechnology for environmental remediation: Materials and applications. Molecules. 2018;**23**: 1760. DOI: 10.3390/molecules23071760

[4] Singh PP, Ambika. 10—Environmental remediation by nanoadsorbents-based polymer nanocomposite. In: Hussain CM, Mishra AK, editors. New Polym. Nanocomposites Environ. Remediat. USA: Elsevier; 2018. pp. 223-241. DOI: 10.1016/B978-0-12-811033-1.00010-X

[5] Khalid S, Shahid M, Niazi NK, Murtaza B, Bibi I, Dumat C. A comparison of technologies for remediation of heavy metal contaminated soils. Journal of Geochemical Exploration. 2017;**182**:247-268. DOI: 10.1016/j.gexplo.2016.11.021

[6] Sharma G, Sharma S, Kumar A, Al-Muhtaseb AH, Naushad M, Ghfar AA, et al. Guar gum and its composites as potential materials for diverse applications: A review. Carbohydrate Polymers. 2018;**199**:534-545. DOI: 10.1016/j.carbpol.2018.07.053

[7] Adekeye DK, Popoola OK, Asaolu SS. Adsorption and conventional technologies for environmental remediation and decontamination of heavy metals: An overview. International Journal of Research and Review. 2019;**6**:505-516

[8] Bushra R. 11—Nanoadsorbents-based polymer nanocomposite for environmental remediation. In: Hussain CM, Mishra AK, editors. New Polym. Nanocomposites Environ. Remediat. USA: Elsevier; 2018. pp. 243-260. DOI: 10.1016/B978-0-12-811033-1.00011-1

[9] Kaur J, Pathak T, Singh A, Kumar K. Application of nanotechnology in the environment biotechnology. In: Kumar R, Sharma AK, Ahluwalia SS, editors. Adv. Environ. Biotechnol. Singapore: Springer; 2017. pp. 155-165. DOI: 10.1007/978-981-10-4041-2_9

[10] Singh V, Yadav P, Mishra V. Recent advances on classification, properties, synthesis, and characterization of nanomaterials. In: Green Synth. Nanomater. Bioenergy Appl. Wiley-Black-Well, Hoboken, New Jersey, USA: John Wiley & Sons, Ltd; 2020. pp. 83-97. DOI: 10.1002/9781119576785.ch3

[11] Baysal A, Kuznek C, Ozcan M. Starch coated titanium dioxide nanoparticles as a challenging sorbent to separate and preconcentrate some heavy metals using graphite furnace atomic absorption spectrometry. International Journal of Environmental Analytical Chemistry. 2018;**98**:45-55. DOI: 10.1080/03067319.2018.1427741

[12] Bakshi M, Abhilash PC. Chapter 17—Nanotechnology for soil remediation: Revitalizing the tarnished resource. In: Singh P, Borthakur A, Mishra PK, Tiwary D, editors. Nano-Mater. Photocatal. Degrad. Environ. Pollut.

USA: Elsevier; 2020. pp. 345-370. DOI: 10.1016/B978-0-12-818598-8.00017-1

[13] Das S, Sen B, Debnath N. Recent trends in nanomaterials applications in environmental monitoring and remediation. Environmental Science and Pollution Research. 2015;**22**:18333-18344. DOI: 10.1007/s11356-015-5491-6

[14] Zhang Y, Wu B, Xu H, Liu H, Wang M, He Y, et al. Nanomaterials-enabled water and wastewater treatment. NanoImpact. 2016;**3-4**:22-39. DOI: 10.1016/j.impact.2016.09.004

[15] Durgalakshmi D, Rajendran S, Naushad M. Current role of nanomaterials in environmental remediation. In: Mu N, Rajendran S, Gracia F, editors. Adv. Nanostructured Mater. Environ. Remediat. Cham: Springer International Publishing; 2019. pp. 1-20. DOI: 10.1007/978-3-030-04477-0_1

[16] Khin MM, Nair AS, Babu VJ, Murugan R, Ramakrishna S. A review on nanomaterials for environmental remediation. Energy & Environmental Science. 2012;**5**:8075-8109. DOI: 10.1039/C2EE21818F

[17] Rawtani D, Khatri N, Tyagi S, Pandey G. Nanotechnology-based recent approaches for sensing and remediation of pesticides. Journal of Environmental Management. 2018;**206**:749-762. DOI: 10.1016/j.jenvman.2017.11.037

[18] Anjum M, Miandad R, Waqas M, Gehany F, Barakat MA. Remediation of wastewater using various nano-materials. Arabian Journal of Chemistry. 2019;**12**:4897-4919. DOI: 10.1016/j.arabjc.2016.10.004

[19] Abd-Elsalam KA. Chapter 1—Multifunctional hybrid nanomaterials for sustainable agri-food and ecosystems: A note from the editor. In: Abd-Elsalam KA, editor. Multifunct. Hybrid Nanomater. Sustain. Agri-Food Ecosyst. USA: Elsevier; 2020. pp. 1-19. DOI: 10.1016/B978-0-12-821354-4.00001-7

[20] Chauhan BPS. Hybrid Nanomaterials: Synthesis, Characterization, and Applications. Hoboken, New Jersey, USA: John Wiley & Sons; 2011

[21] Anaya-Esparza LM, Villagrán-de la Mora Z, Ruvalcaba-Gómez JM, Romero-Toledo R, Sandoval-Contreras T, Aguilera-Aguirre S, et al. Use of titanium dioxide (TiO_2) nanoparticles as reinforcement agent of polysaccharide-based materials. Processes. 2020;**8**:1395. DOI: 10.3390/pr8111395

[22] Meroni D, Ardizzone S. Preparation and application of hybrid nanomaterials. Nanomaterials. 2018;**8**:891. DOI: 10.3390/nano8110891

[23] Rejab MRBM, Hamdan MHBM, Quanjin M, Siregar JP, Bachtiar D, Muchlis Y. Historical development of hybrid materials. In: Hashmi S, Choudhury IA, editors. Encycl. Renew. Sustain. Mater. Oxford: Elsevier; 2020. pp. 445-455. DOI: 10.1016/B978-0-12-803581-8.10546-6

[24] Nguyen T-P, Yang S-H. 19—Hybrid materials based on polymer nanocomposites for environmental applications. In: Jawaid M, Khan MM, editors. Polym.-Based Nanocomposites Energy Environ. Appl. Woodhead Publishing; 2018. pp. 507-551. DOI: 10.1016/B978-0-08-102262-7.00019-2

[25] Vilela C, Pinto RJB, Pinto S, Marques P, Silvestre A, CSdRF B. Polysaccharide Based Hybrid Materials:

Metals and Metal Oxides, Graphene and Carbon Nanotubes. Switzerland: Springer Nature; 2018

[26] Russo T, Fucile P, Giacometti R, Sannino F. Sustainable removal of contaminants by biopolymers: A novel approach for wastewater treatment. Current State and Future Perspectives. Processes. 2021;**9**:719. DOI: 10.3390/pr9040719

[27] Zheng Y, Monty J, Linhardt RJ. Polysaccharide-based nanocomposites and their applications. Carbohydrate Research. 2015;**405**:23-32. DOI: 10.1016/j.carres.2014.07.016

[28] Arora B, Bhatia R, Attri P. 28—Bionanocomposites: Green materials for a sustainable future. In: Hussain CM, Mishra AK, editors. New Polym. Nanocomposites Environ. Remediat. USA: Elsevier; 2018. pp. 699-712. DOI: 10.1016/B978-0-12-811033-1.00027-5

[29] Bilal M, Gul I, Basharat A, Qamar SA. Polysaccharides-based bio-nanostructures and their potential food applications. International Journal of Biological Macromolecules. 2021;**176**:540-557. DOI: 10.1016/j.ijbiomac.2021.02.107

[30] Wen Y, Oh JK. Recent strategies to develop polysaccharide-based nanomaterials for biomedical applications. Macromolecular Rapid Communications. 2014;**35**:1819-1832. DOI: 10.1002/marc.201400406

[31] Kotharangannagari VK, Krishnan K. Biodegradable hybrid nanocomposites of starch/lysine and ZnO nanoparticles with shape memory properties. Materials and Design. 2016;**109**:590-595. DOI: 10.1016/j.matdes.2016.07.046

[32] Nasrollahzadeh M, Sajjadi M, Iravani S, Varma RS. Starch, cellulose, pectin, gum, alginate, chitin and chitosan derived (nano)materials for sustainable water treatment: A review. Carbohydrate Polymers. 2021;**251**: 116986. DOI: 10.1016/j.carbpol.2020.116986

[33] Robyt JF. Starch: Structure, properties, chemistry, and enzymology. In: Fraser-Reid BO, Tatsuta K, Thiem J, editors. Glycosci. Chem. Chem. Biol. Berlin, Heidelberg: Springer; 2008. pp. 1437-1472. DOI: 10.1007/978-3-540-30429-6_35

[34] Campelo PH, Sant'Ana AS, Pedrosa Silva Clerici MT. Starch nanoparticles: Production methods, structure, and properties for food applications. Current Opinion in Food Science. 2020;**33**:136-140. DOI: 10.1016/j.cofs.2020.04.007

[35] Kim H-Y, Park SS, Lim S-T. Preparation, characterization and utilization of starch nanoparticles. Colloids and Surfaces. B, Biointerfaces. 2015;**126**:607-620. DOI: 10.1016/j.colsurfb.2014.11.011

[36] Yu M, Ji N, Wang Y, Dai L, Xiong L, Sun Q. Starch-based nanoparticles: Stimuli responsiveness, toxicity, and interactions with food components. Comprehensive Reviews in Food Science and Food Safety. 2021;**20**:1075-1100. DOI: 10.1111/1541-4337.12677

[37] Ogunsona E, Ojogbo E, Mekonnen T. Advanced material applications of starch and its derivatives. European Polymer Journal. 2018;**108**:570-581. DOI: 10.1016/j.eurpolymj.2018.09.039

[38] Ragab E, Shaban M, Khalek AA, Mohamed F. Design and characterization of PANI/starch/Fe_2O_3 bio composite for wastewater remediation. International Journal of Biological Macromolecules. 2021;**181**:301-312. DOI: 10.1016/j.ijbiomac.2021.03.043

[39] Dehabadi L, Wilson LD. Polysaccharide-based materials and their adsorption properties in aqueous solution. Carbohydrate Polymers. 2014;**113**:471-479. DOI: 10.1016/j.carbpol.2014.06.083

[40] Majhi KC, Yadav M. Chapter 5—Synthesis of inorganic nanomaterials using carbohydrates. In: Inamuddin BR, Ahamed MI, Asiri AM, editors. Green Sustain. Process Chem. Environ. Eng. Sci. USA: Elsevier; 2021. pp. 109-135. DOI: 10.1016/B978-0-12-821887-7.00003-3

[41] Corsi I, Fiorati A, Grassi G, Pedrazzo AR, Caldera F, Trotta F, et al. Chapter 14—Ecosafe nanomaterials for environmental remediation. In: Bonelli B, Freyria FS, Rossetti I, Sethi R, editors. Nanomater. Detect. Remov. Wastewater Pollut. USA: Elsevier; 2020. pp. 383-405. DOI: 10.1016/B978-0-12-818489-9.00014-1

[42] Khademian E, Salehi E, Sanaeepur H, Galiano F, Figoli A. A systematic review on carbohydrate biopolymers for adsorptive remediation of copper ions from aqueous environments-part A: Classification and modification strategies. Science of the Total Environment. 2020;**738**:139829. DOI: 10.1016/j.scitotenv.2020.139829

[43] Le Corre D, Angellier-Coussy H. Preparation and application of starch nanoparticles for nanocomposites: A review. Reactive and Functional Polymers. 2014;**85**:97-120. DOI: 10.1016/j.reactfunctpolym.2014.09.020

[44] Pooresmaeil M, Namazi H. Chapter 14—Application of polysaccharide-based hydrogels for water treatments. In: Chen Y, editor. Hydrogels Based Nat. Polym. Elsevier; 2020. pp. 411-455. DOI: 10.1016/B978-0-12-816421-1.00014-8

[45] Zubair M, Jarrah N, Ihsanullah KA, Manzar MS, Kazeem TS, et al. Starch-NiFe-layered double hydroxide composites: Efficient removal of methyl orange from aqueous phase. Journal of Molecular Liquids. 2018;**249**:254-264. DOI: 10.1016/j.molliq.2017.11.022

[46] Babaee Y, Mulligan CN, Rahaman MS. Stabilization of Fe/Cu nanoparticles by starch and efficiency of arsenic adsorption from aqueous solutions. Environmental Earth Sciences. 2017;**76**:650. DOI: 10.1007/s12665-017-6992-z

[47] Gong Y, Wang L, Liu J, Tang J, Zhao D. Removal of aqueous perfluorooctanoic acid (PFOA) using starch-stabilized magnetite nanoparticles. Science of the Total Environment. 2016;**562**:191-200. DOI: 10.1016/j.scitotenv.2016.03.100

[48] Okuo J, Emina A, Omorogbe S, Anegbe B. Synthesis, characterization and application of starch stabilized zerovalent iron nanoparticles in the remediation of Pb-acid battery soil. Environmental Nanotechnology, Monitoring and Management. 2018;**9**:12-17. DOI: 10.1016/j.enmm.2017.11.004

[49] Mosaferi M, Nemati S, Khataee A, Nasseri S, Hashemi AA. Removal of Arsenic (III, V) from aqueous solution by nanoscale zero-valent iron stabilized with starch and carboxymethyl cellulose. Journal of Environmental Health Science and Engineering. 2014;**12**:74. DOI: 10.1186/2052-336X-12-74

[50] Dong H, He Q, Zeng G, Tang L, Zhang C, Xie Y, et al. Chromate removal by surface-modified nanoscale zero-valent iron: Effect of different surface coatings and water chemistry. Journal of Colloid and Interface Science. 2016;**471**:7-13. DOI: 10.1016/j.jcis.2016.03.011

[51] Naushad M, Ahamad T, Sharma G, Al-Muhtaseb AH, Albadarin AB, Alam MM, et al. Synthesis and characterization of a new starch/SnO_2 nanocomposite for efficient adsorption of toxic Hg^{2+} metal ion. Chemical Engineering Journal. 2016;**300**:306-316. DOI: 10.1016/j.cej.2016.04.084

[52] Chen Y, Zhao W, Wang H, Meng X, Zhang L. A novel polyamine-type starch/glycidyl methacrylate copolymer for adsorption of Pb(II), Cu(II), Cd(II) and Cr(III) ions from aqueous solutions. Royal Society Open Science. 2018;**5**:180281. DOI: 10.1098/rsos.180281

[53] Rashid US, Simsek S, Kanel SR, Bezbaruah AN. Modified tapioca starch for iron nanoparticle dispersion in aqueous media: Potential uses for environmental remediation. SN Applied Sciences. 2019;**1**:1379. DOI: 10.1007/s42452-019-1364-9

[54] Moradi E, Ebrahimzadeh H, Mehrani Z, Asgharinezhad AA. The efficient removal of methylene blue from water samples using three-dimensional poly (vinyl alcohol)/starch nanofiber membrane as a green nanosorbent. Environmental Science and Pollution Research 2019;**26**:35071-35081. https://doi.org/10.1007/s11356-019-06400-7.

[55] Xia K, Liu X, Wang W, Yang X, Zhang X. Synthesis of modified starch/polyvinyl alcohol composite for treating textile wastewater. Polymers. 2020;**12**:289. DOI: 10.3390/polym12020289

[56] Chowdhury MNK, Ismail AF, Beg MDH, Hegde G, Gohari RJ. Polyvinyl alcohol/polysaccharide hydrogel graft materials for arsenic and heavy metal removal. New Journal of Chemistry. 2015;**39**:5823-5832. DOI: 10.1039/C5NJ00509D

[57] Sharma G, Naushad M, Kumar A, Rana S, Sharma S, Bhatnagar A, et al. Efficient removal of coomassie brilliant blue R-250 dye using starch/poly(alginic acid-cl-acrylamide) nanohydrogel. Process Safety and Environment Protection. 2017;**109**:301-310. DOI: 10.1016/j.psep.2017.04.011

[58] Nsom MV, Etape EP, Tendo JF, Namond BV, Chongwain PT, Yufanyi MD, et al. A green and facile approach for synthesis of starch-pectin magnetite nanoparticles and application by removal of methylene blue from textile effluent. Journal of Nanomaterials. 2019;**2019**:e4576135. DOI: 10.1155/2019/4576135

[59] Robinson MR, Coustel R, Abdelmoula M, Mallet M. As(V) and As(III) sequestration by starch functionalized magnetite nanoparticles: Influence of the synthesis route onto the trapping efficiency. Science and Technology of Advanced Materials. 2020;**21**:524-539. DOI: 10.1080/14686996.2020.1782714

[60] Chang PR, Zheng P, Liu B, Anderson DP, Yu J, Ma X. Characterization of magnetic soluble starch-functionalized carbon nanotubes and its application for the adsorption of the dyes. Journal of Hazardous Materials. 2011;**186**:2144-2150. DOI: 10.1016/j.jhazmat.2010.12.119

[61] Chen Y, Guo Z, Das R, Jiang Q. Starch-based carbon nanotubes and graphene: Preparation, properties and applications. ES Food & Agroforestry. 2020;**2**:13-21. DOI: 10.30919/esfaf1111

[62] Bhuyan MM, Chandra Dafader N, Hara K, Okabe H, Hidaka Y, Rahman MM, et al. Synthesis of potato starch-acrylic-acid hydrogels by gamma radiation and their application in dye adsorption. International Journal of

Polymer Science. 2016;**2016**:e9867859. DOI: 10.1155/2016/9867859

[63] Gomes RF, de Azevedo ACN, Pereira AGB, Muniz EC, Fajardo AR, Rodrigues FHA. Fast dye removal from water by starch-based nanocomposites. Journal of Colloid and Interface Science. 2015;**454**:200-209. DOI: 10.1016/j.jcis.2015.05.026

[64] Gupta R, Xie H. Nanoparticles in daily life: Applications, toxicity and regulations. Journal of Environmental Pathology, Toxicology and Oncology. 2018;**37**(3):209-230. DOI: 10.1615/JEnvironPatholToxicolOncol.2018026009

[65] Kasai T, Umeda Y, Ohnishi M, Kondo H, Takeuchi T, Aiso S, et al. Thirteen-week study of toxicity of fiber-like multi-walled carbon nanotubes with whole-body inhalation exposure in rats. Nanotoxicology. 2015;**9**:413-422. DOI: 10.3109/17435390.2014.933903

[66] Ray PC, Yu H, Fu PP. Toxicity and environmental risks of nanomaterials: Challenges and future needs. Journal of Environmental Science and Health, Part C: Environmental Carcinogenesis & Ecotoxicology Reviews. 2009;**27**:1-35. DOI: 10.1080/10590500802708267

[67] Xie F, Pollet E, Halley PJ, Avérous L. Advanced nano-biocomposites based on starch. In: Ramawat KG, Mérillon J-M, editors. Polysacch. Bioactivity Biotechnol. Cham: Springer International Publishing; 2015. pp. 1467-1553. DOI: 10.1007/978-3-319-16298-0_50

[68] Villaseñor MJ, Ríos Á. Nanomaterials for water cleaning and desalination, energy production, disinfection, agriculture and green chemistry. Environmental Chemistry Letters. 2018;**16**:11-34. DOI: 10.1007/s10311-017-0656-9

[69] Sun Y, Jing R, Zheng F, Zhang S, Jiao W, Wang F. Evaluating phytotoxicity of bare and starch-stabilized zero-valent iron nanoparticles in mung bean. Chemosphere. 2019;**236**:124336. DOI: 10.1016/j.chemosphere.2019.07.067

[70] Ciambelli P, La Guardia G, Vitale L. Chapter 7—Nanotechnology for green materials and processes. In: Basile A, Centi G, Falco MD, Iaquaniello G, editors. Stud. Surf. Sci. Catal. Vol. 179. USA: Elsevier; 2020. pp. 97-116. DOI: 10.1016/B978-0-444-64337-7.00007-0